KIDSOURCE

SCIENCE FAIR
HANDBOOK

By Danna Voth
Illustrated by Mike Moran

LOWELL HOUSE JUVENILE

LOS ANGELES

NTC/Contemporary Publishing Group

Library of Congress Cataloging-in-Publication Data

```
Voth, Danna.
    Kidsource : science fair handbook / by Danna Voth.
        p.    cm.
    Includes bibliographical references and index.
    Summary: Provides information on choosing and planning a science
fair project, finding and using resources, applying scientific
principles, and delivering the final project.
    ISBN 1-56565-514-1 (pbk.)
    1. Science--Exhibitions--Handbooks, manuals, etc.--Juvenile
literature.   [1. Science projects--Methodology.]   I. Title.
Q105.V68   1998
507.8--dc21
                                                          98-25059
                                                             CIP
                                                             AC
```

Published by Lowell House
A division of NTC/Contemporary Publishing Group, Inc.
4255 West Touhy Avenue, Lincolnwood (Chicago), Illinois 60646-1975 U.S.A.

Managing Director and Publisher: Jack Artenstein
Director of Publishing Services: Rena Copperman
Editorial Director, Juvenile: Brenda Pope-Ostrow
Director of Juvenile Development: Amy Downing
Director of Art Production: Bret Perry
Designer: Treesha Runnells

Lowell House books can be purchased at special discounts when ordered in bulk for
premiums and special sales. Contact Customer Service at the above address, or call:
1-800-323-4900

Printed and bound in the United States of America

10 9 8 7 6 5 4 3 2 1

CONTENTS

ACKNOWLEDGMENTS

I would like to thank Jim Hastings, Dr. Arnie Korporaal, and Darrell W. Smedley, of the Los Angeles County Office of Education, for their enormous support and the wonderful job they do with the annual Los Angeles County Science Fair. Linda Cady, a dynamite science teacher at science magnet school John Adams, offered hours of her time, indispensible wisdom, and armloads of materials. Another fabulous science teacher, Nancy Bernstein, proved inspiring as she shared her creative approach to teaching. Dr. E. C. Krupp, a famous archeoastronomer at the Griffith Observatory, was very generous with his time and is the sort of knowledgeable, funny, and kind expert I hope children may always be fortunate enough to contact.

I would like to thank my family and good friends for their support during this project: Thank you, Donna Rutledge, Valerie Voth, Julie Voth, Laura Shamas, Morgan Delancey, Michele Melden, Tom Haun, Robert Arnold, Sally Gardea, Linda Campo, Sherrie Murphey, Marcia Walker, Janet Walker, John Gootherts, Tina Wall, and Carol Woodliff.

This book could never have happened without the wonderful students who do science projects all across the country, and I want to especially thank the following students for their time and enthusiasm: Brian Bourget, Dara Weinberg, Gerald Larue, Khalil Sharif, Linda Day, Keesha Johnson, Bettina Kwan, Maya Plinke, Raphael Navarro, and Zephyr Detrano. Also thank you to the parents and teachers, particularly Debra Bourget, a great and encouraging mom, who helped me arrange to meet and talk with the students.

And finally, I want to thank my editor at Lowell House, Amy Downing, whose interest and encouragement in this project made it a reality.

INTRODUCTION

What could be more exciting than learning science? Doing science! Now you have a chance to do your own science project and show your results at a science fair. Sound just a little scary? Whether this is your first science fair or your fifth, *KidSource* can help you make this project your best!

First, *KidSource: Science Fair Handbook* gives you the basics on science fair projects and procedures and explains how to work scientifically.

Next, *KidSource* will help you create the perfect project for you—one that is original and exciting!

To help you prepare for your project, *KidSource*'s Information Director explains step-by-step how to do research—from using your library to contacting people who can help you. Need help finding supplies? The Information Director also tells you where to look for everything you may need.

KidSource then takes you through running your experiment to creating a dynamite display for the fair, with instructions on how to build your exhibit and how to make it special. You will find important timesaving tips and a timetable to help you get everything done on time. And if your science fair is a competition, *KidSource* helps you prepare for the judges.

Finally, *KidSource* provides you with a handy chapter on the tools of science and a list of books and magazines for further reading. In addition, you'll find the **bold** words throughout defined in the Glossary at the back.

KidSource: Science Fair Handbook is here to help you every step of the way, from coming up with an idea to the final day of the fair!

WHAT IS A SCIENCE FAIR?

"I started out just collecting rocks. But then I had to find out about them. It was like going back in time. I can tell you four different stories about what happened in my backyard, millions of years ago. So my project was like a time machine."

—Maya Plinke, age 11

So your school is having a science fair, and you want to participate. But wait! What is a science fair? That's the place you and your classmates take your science project exhibits to show all the hard work you have done.

Great! Now, what's a science project? And what is a science project exhibit?

Your science project is the experiment or field study you do to explore a question you thought up.

Your science project exhibit is the display you build to show the steps and the results of your science project. That's what you bring to the fair!

Each science fair has its own special rules to let you know how big your exhibit can be, what materials you may use, and what types of experiments you may do.

Usually the fair has categories for subjects of experiments, different divisions based on student ages or grade levels, and individual or team work. Here's some more info about science fairs.

HISTORY

U.S. science fairs began in 1928, when the American Institute of New York, which was organized to promote domestic industry, sponsored the first Children's Fair. Soon science clubs were formed, and science fairs became so popular that the American Institute could not support all the interest in science fairs by itself. Science Service of Washington, D.C., a nonprofit institution dedicated to making science popular, took over the national effort to develop science fairs for students. All across the United States, numerous science fairs were held, and chemical, optical, engineering, and pharmaceutical companies, as well as individuals, community groups, and government agencies, joined in the effort to keep the fairs going and growing. In 1997, over 500 fairs nationwide were directly associated with Science Service. But those are just the tip of the iceberg. Many high schools hold their own science fairs and send their top contestants to Science Service–sponsored science fairs, hoping to win national recognition. Also, plenty of science fairs are held at elementary and junior high schools across the country.

WORK YOUR TV!

Here's an idea to explore if you are interested in how people behave. The study of how people in organized groups behave is called **sociology.**

Did you know people behave differently when they are being watched than when they are alone? The television program *Candid Camera* showed what funny things people did when they thought they were alone. How do you think people will act when they know that others may be looking at them? Try watching crowd scenes from television coverage of sporting events, theater presentations, and public meetings, such as a congressional debate or a city council meeting. Keep a record of each scene you watch and how people behave. Did your survey show what you expected, or did it surprise you?

WHERE

Some science fairs are small and may be organized by parents and the science teachers at your school. These are usually held in a gym, auditorium, or classroom. Others, with many schools participating, are larger and take place in convention centers, public halls, or indoor arenas big enough to arrange many rows of tables where students can place their exhibits.

WHEN

Science fairs are held in the spring and usually last about three or four days. On the first day you set up your exhibit. It may be on display for a few days so family, friends, and classmates can see it, along with all the projects. Some science fairs are presented primarily to honor and encourage student participation, and some science fairs are competitive. If your science fair is a competition, once your exhibit is set up, judges look at it and the other exhibits and evaluate them, and on the next day you may be invited to talk with the judges and tell them about your project. Then prizes are awarded, and the public is encouraged to come and see all the exhibits. After that, you take your exhibit home. If the fair is small, these things may happen a little faster, and if the fair is large, more time may be needed.

KIDSOURCE TIPS

You will learn a lot about what makes a good exhibit if you visit a science fair before you actually enter your project in one.

Find out when the students will be at the fair, and talk to them about their projects. Some good questions are: What would you do differently next time? How long did it take you to prepare your project? Who or what did you find most helpful during the process?

So, even though you may not be doing a science project just yet, ask your teacher about upcoming science fairs in your area.

WHO'S WHO

It takes a lot of people to make a science fair happen: the participants, the organizers of the fair, and the judges who evaluate the exhibits.

Participants—like you—are students from one or more schools and are usually grouped by age or grade.

The people who organize the fair may be your teachers, or parents. Sometimes people from private companies, your school district, your county board of education, or your state's universities will get involved.

Judges are teachers, doctors, veterinarians, engineers, psychologists, geologists, chemists, physicists, biologists, and professional government workers and business people who have expertise in an area of science or work in an industry that uses scientific knowledge.

WHY

Why are you doing a science project and participating in a fair? Sometimes you do a project because it is a school assignment, but here are lots of other smart reasons to participate in a science fair:

- to learn how to think scientifically and see what it's like to be a scientist
- to make your own observations about the world
- to talk with professional scientists
- to work with people in your community

- to learn how to organize and finish a long-term project

- to learn to present your work and ideas in a clear and interesting way

- to add to the world's knowledge with your **data**

- to develop new skills, like researching, measuring, and calculating

- to build and operate scientific equipment

Student Experience

"With my first science fair project, I started to see how strange things can be. I was trying to find out how to make wildflower seeds grow, and I discovered they had all these weird abilities that matched their environment, like the seeds that only sprouted if they had been scraped, or soaked, or burned. It was awesome!"

—Keesha Johnson, age 12

- to help improve the world through your project's results.

HELP IS HERE

It takes a lot of planning and effort to make it to the science fair, and your teachers, family, friends, and professional community can help you.

Teachers are great. Your science teacher can answer lots of questions, help you look for ideas, order supplies, check your organization, and show you ways to find out more about your topic. Your shop teacher can help you build your exhibit and any equipment you may need to construct. Your art teacher can help you make your exhibit eye-catching. Your English teacher can help you write your report and your bibliography and give you research tips. Your math teacher can check your calculations and help you with charts and graphs.

Your parents or guardian can help you do your project safely, assist with transportation and technical things like construction or photography, and get supplies. They can share ideas and their own experience as well as be interested listeners. You can practice talking about your project with your friends and family, so you will feel more comfortable during your interview with the judges.

The professional community can also be a great help. Contact the authors of the most interesting articles from your research to ask them detailed questions. You also can speak with scientists who work at museums, laboratories, and universities about your experiment and your data.

SCIENCE FAIR GUIDELINES

Science fair rules will vary from fair to fair, so always make sure you receive a list of guidelines. Here are some guidelines used at most science fairs:

- Your display must be self-supporting, which means it must be stable enough to stand up on a table without being taped or otherwise connected to the table. It usually should be no more than 4 feet wide, 6½ feet tall, and 2½ feet deep.

- Any electrical devices in your display must not have any exposed wires or connections. Hot parts such as lightbulbs must be shielded out of reach of viewers.

- No dangerous chemicals, fires, **toxins,** or combustible materials may be used in your display.

- Only one project is allowed per student, and all work must be your own, unless you are working on a team.

- You must be available to speak to the judges on the day and time assigned to you.

Designer Pasta

You don't always need expensive materials to make a great science project—just imagination.

Car designers often work with models and test out their ideas on them before building more expensive versions. For the 1997 Los Angeles County Science Fair, Jenny Ly, from West Covina, California, created an automobile engineering project that compared various combinations of length and mass for speed. To test her models, she held a pastamobile race—she built her cars out of pasta!

DO YOU WANT TO USE ANIMALS IN YOUR PROJECT?

First, think about why you want to use animals. Is your purpose humane—will it benefit animals in general, or the animals you will involve in your project?

Also, think about the responsibility of working with animals. Do you know how to care for your subject animals properly? If you are going to keep animals for your project, you must have a safe place where you can work with them. What do they eat and drink, and how often? What kind of space and how much room do they need?

Are you prepared to spend time and money to take care of them, even when you'd rather be riding your bike or spending your money on a new computer game? Are you sure you want to clean up after them? Can you afford veterinary care, if necessary? Can you find someone to take care of your animals if your family goes on a vacation? And most important: What will happen to your animals after your experiment is over?

• Do not bring to the fair any live animals, preserved animal or human body parts, fungi or microbial cultures, hypodermic needles, pipettes, drugs, open-top battery cells, explosive or radioactive materials, or anything with exposed sharp edges. While you may be allowed to use some of these things in your experiment, you still may not bring them to the fair to display in your exhibit.

- You must fill out special forms and follow the specific rules they describe if you plan to use tissue samples, animals, or human subjects in your experiment. These forms must be turned in and OK'd before you begin your experiment. You may not get tissue samples yourself but must order them from a qualified biological supplier.

- You must treat animals and human subjects responsibly and safely. You may not hurt, drug, or subject animals to any kind of stress, or embarrass or put people at any kind of risk, emotional or physical. All human subjects' rights to privacy must be respected.

Keeping Your Subjects' Privacy Safe

If you use human subjects in your experiment, they may want to remain anonymous.

What's a good way to respect the privacy of the people you use in your project? Make two lists on one sheet of paper. On one list, write each subject's name, and on the other list, assign each a number. Do this for all your subjects. Put this paper somewhere for safekeeping, and identify your

MADELINE HEDRICK 1
CHLOE MARES 2
MILES RIVERA 3
NICHOLAS STEVENS 4
JACOB GULLETT 5
MEGAN PEREZ 6
VIN GALLO 7
LEWIS MOORES 8

subjects in all your reports only by the numbers you have assigned them. Save the list of your subjects' names in case you need to ask them any questions later. Then you can look up the number to find the corresponding name.

SCIENCE FAIR BASICS

"This was my first project, and I was learning how to do a project as much as I was learning about how plants grow."

—Linda Day, age 13

Science is based on wondering. You begin to be a scientist when you ask questions:

- Why did that happen?

- What would be different if I changed this one thing?

- How did that happen?

- When did that occur?

- How is this different from that?

You become a scientist when you try to find answers to your questions by using the scientific method.

When you follow the scientific method, your science project begins with a **hypothesis**—a question and your own informed guess at an answer, which you **test** by following your **procedure**. A procedure is the steps you take to do an **experiment** or **field work,** which leads you to confirm—or not confirm—your hypothesis. You look at the actual results, compare them with your expectations, and write your **conclusion** based on what you have found out.

In your **report,** you describe how you followed the scientific method, step by step. At the end of your report, you will mention new questions you would like to look into and things you would like to try based on what you have learned from your results. That's what makes science projects so fun—they can lead to lots of new discoveries!

THE BIRTH OF THE SCIENTIFIC METHOD

Galileo Galilei (1564–1642), an Italian astronomer and physicist, believed in discovering facts by first forming a theory (or hypothesis) and then testing it in an experiment, just as you do in your science project. This approach, which we call the scientific method, was a very radical idea in his time. People were more inclined to accept ideas because they agreed with their religious beliefs or, to them, seemed logical. But many scientific discoveries do not appear at first to make sense. For instance, because they could not see something with their naked eyes, some of Galileo's contemporaries did not believe that other planets had satellites. Galileo also believed in careful observation and measurement and developed the telescope into a powerful tool for exploring the sky. With this tool he discovered four of Jupiter's satellites. Galileo's desire to test his theories through experiment and learn through observation and measurement moved science forward and offers a good model for us today.

THE SCIENTIFIC METHOD

Let's take some time to understand the scientific method, the backbone of your science fair project. The scientific method has four parts:

OBSERVATION

You notice something in the world that you want to know more about. You then ask a question about it. This question is what you try to uncover an answer to in your science fair project.

Jumping Genes

All living things have genes, and scientists have been trying to unlock the secret of the genetic code since the discovery of DNA. Here is the story of an exciting piece of the puzzle that was found by a dedicated scientist. It shows how important careful observation is to science.

Barbara McClintock, who won the Nobel prize in physiology/medicine in 1983, made her important discovery of "jumping genes" by being very observant. She was studying maize (corn) chromosomes—a chromosome being the part of the cell where the genetic code, DNA, is stored. Because of her careful observation, she noticed a change in cell coloring in the corn's leaves. She decided to look at the plant's chromosomes before and after the coloring change occurred—she guessed that a change, or mutation, in the chromosomes had caused the change in coloring.

Because the cells of maize are very large and the new growth of maize is easy to track, McClintock was able to view the cell chromosomes before and after the coloring change occurred. She found that some genes had moved around, and when they moved, they sometimes dropped or took adjacent bits of DNA with them, causing mutations. She had discovered jumping genes!

HYPOTHESIS

You **predict** why, when, where, or how whatever you observed happened, based on information you already have. Sometimes this takes the form of an "if . . . then" statement. A hypothesis is often called an "educated guess" because you base your prediction on facts you already know.

TESTING

You test your hypothesis with a procedure. You can do either an experiment, where everything except the particular thing being tested is carefully controlled, or field work, where you study your subject in the natural world. Careful observations and measurements are recorded in both testing procedures.

CONCLUSION

You state whether or not your hypothesis was correct, based on the results of your testing. If your hypothesis is proven wrong, try to explain why. Also, make any further predictions your results could point to, and describe any changes to your procedure you think would give more accurate results or be helpful to further research.

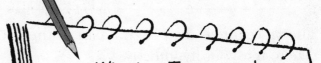

Winning Teamwork

"Working in a team makes it easier to go in depth, to do more than we could have done on our own. I enjoyed it a lot."
　　—Gerald Larue, seventh-grader

Sometimes science fairs allow team entries. Having a partner can mean a lot of help with the planning and work of a project. Partners can split the steps of the procedure, as seventh-graders Khalil Sharif and Gerald Larue did. One student gathered samples, and the other tested them. "It's a lot of work, so we divided some things and worked on some things together and combined our skills," says Khalil, who, with partner Gerald, won third prize at the 1996 California State Science Fair for their team project, "The Effect of Storm Run-Off on Water Characteristics of the Los Angeles Harbor."

PROCEDURE

The procedure is the practical part of the scientific method—it's the steps you take to test your hypothesis.

The purpose of science is to discover things about the world, with accuracy, truth, and objectivity. Scientists

- test ideas

- weigh evidence carefully

- come to conclusions cautiously

- make conclusions based on facts.

An important part of the scientist's process of discovery is the procedure followed. A procedure is like a recipe—it's a list of steps. The steps you plan to take to test your hypothesis must be clearly written out so that anyone could repeat what you have done. Your procedure

- gives step-by-step directions on what to do

- lists all the materials and equipment you use

- provides any instructions you need to build or use equipment.

EXPERIMENT AND FIELD WORK

Scientists test their hypotheses either through experiment or field work.

EXPERIMENT

Experimental observations are made in a controlled **environment** that you create. How? You make a simpler, smaller-scale version of the part of the real world you want to study. You focus your attention on just a few things, instead of on everything, that can happen.

In an experiment, a scientist tries to look at how just one thing affects a subject. The tricky part is creating an environment in which only that one thing changes. That is why you often see scientists using test tubes, petri dishes, and other small, enclosed settings for their experiment. It is much easier to control things in such environments.

FIELD WORK

In field work, a scientist goes into an uncontrolled environment, a specific place in the world, and records exactly what is observed there at the time. Because you are studying a unique situation every time, field work is almost always new and original.

The tricky part with field work is that while you are recording your observations, you must make sure that you yourself are not interfering with your subject simply by being there. For instance, you cannot count birds in a tree if you scare any (or all!) away while you try to count them.

FIELD STUDY FINDS NEW LIFE-FORMS

Your field study may occur in your own backyard or even at the local nature preserve, but can you imagine exploring the deep rifts in the ocean's floor in a submarine?

In 1977, scientists aboard the research submarine *Alvin,* from Woods Hole Oceanographic Institute, discovered a new ecosystem, or community of organisms, thriving near volcanic vents at the bottom of the freezing waters of the Pacific Ocean's Galapagos Rift.

The high heat and hydrogen sulfide from the cracks in the volcanoes provide the energy for special bacteria, a staple of the unique food chain there. Other members of the ecosystem are huge tube worms up to 25 feet long. Because they are so unlike anything known, they are classified in a phylum, Vestimentifera, by themselves. The scientists named some of the new worms alvinellid worms, after their submarine research vessel.

This strange volcanic ecosystem, based on converting sulphurous chemicals into food, suggests to some scientists the possibility that there may be similar strange life-forms on other planets with volcanic activity.

VARIABLES, CONTROLS, GROUPS, AND TRIALS

Scientists are like detectives—they try to solve mysteries. Experiments are part of a scientist's detective kit. When you want to prove a theory true or false, create an experiment that will test one thing you can observe.

You have an idea—that if you set up a controlled situation and purposely change only one thing, this alteration will cause something else to happen. The thing you purposely change is called the **changing variable**. If your change causes something else to happen, this "something else" is called the **responding variable,** because it is responding to the change.

You must plan your procedure carefully to be sure that you change only one thing in your **experimental group**.

Suppose you want to know what would happen if you played music for an experimental group of plants. You will play music, your changing variable, and watch for any signs of a responding variable, which you expect to be bigger or faster growth.

But how will you know if any growth is a change? How will you know what is bigger and faster growth? You need a way to compare the rate of growth. You need to have something to compare your experimental plants to—something to show what normal growth is. So you need a **control group**. You need to raise some other plants in exactly the same way as you raise your experimental group, except that they will not experience the changing variable. You will treat them exactly the same as you do the experimental group, but you will not play music for them.

Your experimental group—Give these plants x amount of food, y amount of water, and play music for them.

Your control group—Use the same kind and age and size of plants, give them the same amounts of food and water, but do not expose them to any music.

You can measure the growth of the plants that you expose to music against the growth of the plants that you don't.

Still, you need to consider some other things. Can you think of anything that could affect the plants? How about diseases or pests? Could some of the plants have been healthier than others before you even started the experiment? That is possible, even though you looked them over carefully before you began. To ensure that any recorded change is from your changing variable, and only from your changing variable, you should test in groups of at least 25 subjects.

For example, if you only tested one or two plants and they both died, you could not be sure that their death resulted from your experiment, or if they were weak before being part of the experiment and were about to die anyway. But if you tested a group of 25 subjects and only two died, you could more confidently conclude that those two plants had been weak or ill before the experiment began.

To be reasonably sure that nothing happens randomly (by chance), you also should run at least three **trials**—do your experiment three times. For example, if you ran your experiment once and correctly used a group of at least 25 subjects, but they all died, you could not be sure their death was the result of your experiment. Perhaps they had all been weak or ill before they were affected by your experiment. If you have at least three trials, and the results are similar each time, you can feel more confident that your results are accurate. If one of the trials gives results that are inconsistent with the others, you can suspect a problem with the odd trial.

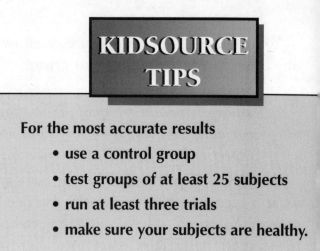

KIDSOURCE TIPS

For the most accurate results
- **use a control group**
- **test groups of at least 25 subjects**
- **run at least three trials**
- **make sure your subjects are healthy.**

LET'S SEE!

Here are some examples of variables and control groups. In Table 1, take the subject and apply the changing variable to it to see if there would be a change in the responding variable. Table 2 adds a control group that does not experience the changing variable.

If change takes place in the experimental group but not in the control group, you can be reasonably sure your changing variable caused the alteration in the responding variable.

If change occurs in both the experimental and control groups, something else, other than your changing variable, caused the alteration in the responding variable.

Here are three hypotheses to test:

- If I add alcohol to water, the freezing time will change.

- If I give plants plant food, the rate of growth will change.

• If I add color to potatoes, the taste will change.

Here are variables to test the hypotheses in three different experiments:

TABLE 1 VARIABLES

Subject	Changing Variables	Responding Variables
WATER	ALCOHOL	THE FREEZING TIME
PLANTS	PLANT FOOD	THE RATE OF GROWTH
POTATO	COLOR	THE TASTE OF THE POTATO

Here are experimental and control groups to test in the three different experiments.

TABLE 2 EXPERIMENTAL & CONTROL GROUPS

Subject	Experimental Group	Control Group
WATER	ALCOHOL ADDED	NO ALCOHOL ADDED
PLANTS	PLANTS GETTING FOOD	GETTING NO FOOD
POTATO	PEOPLE TASTE POTATO WITH NO COLOR ADDED	SAME PEOPLE TASTE POTATO WITH COLOR ADDED

OBSERVATIONS AND MEASUREMENTS

Every time you look at something, listen to something, or feel something, you are actually measuring it. Some of our most important discoveries, such as the invention of the microscope, telescope, and thermometer, have been finding ways to help us measure things. You can measure things with scales, thermometers, barometers, microscopes, telescopes, spectroscopes, and geiger counters.

MICROSCOPE

It is very important that your measurements, and the words you use to describe your measurements, be exact. What is the difference between "less than a liter" and "almost a liter"? If you tell someone to turn "a little way" down the road, couldn't they make a wrong turn and get lost? In order to get accurate results in science, you need to be exact when you measure and when you describe things. One reason for this is that anyone else should be able to follow your procedure and get the same results as you.

BAROMETER

SPECTROSCOPE

TELESCOPE

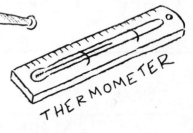

THERMOMETER

AMPLIFIER

GEIGER COUNTER

Measures of Man

Early people developed systems of measurement for building and for trading fairly. They began by using the human body as their unit of measure for distances. The lengths of fingers, palms, feet, and strides were used.

The Egyptians used a complex but incredibly accurate system based on palms to build their pyramids, which usually deviated only 5 inches over a span of 700 feet.

The ancient Babylonians developed a system of weights based on metal objects of various related sizes and weights. They had the mina, a duck-shaped trinket, and the swan, weighing 30 minas.

a mina

QUALITATIVE AND QUANTITATIVE MEASUREMENTS

There are two ways to observe things: qualitatively and quantitatively. **Qualitative data** describes the quality, or characteristic, of something. If something grows, boils, freezes, burns, tastes sour, changes color, dries out, smells, moves, makes noise, or changes in any way from its original state, you want to make note of that. You should describe shapes, colors, smells, sounds, activity, and any characteristic you can.

Usually you can find out more about what is happening, and then describe it in greater detail by measuring it with a standard measuring tool such as a ruler or scale. That is finding **quantitative data**.

When you want to describe a qualitative measurement more precisely, you make a quantitative measurement. You may have described a sample as "very cold." But if you want a more precise description, you'll want to measure the temperature of your sample with a thermometer.

The most important thing in quantitative measurement is knowing how to use the standard measuring tool you have. When you measure the length of something with a ruler, you cannot say you know exactly how long something is but only that you know the length plus or minus the smallest part marked on the ruler. That is because you are only guessing the distance between the smallest markings. If the smallest lines are millimeters, then your measurement is plus or minus 1 millimeter. When you stand on a scale in the bathroom, it usually only marks the pounds, and so you estimate the partial pounds, the area indicated between the lines. The scale measurement is plus or minus 1 pound.

When you are measuring liquids, you may be using a graduated cylinder, measuring cups, eyedroppers, or other devices. There is a special trick to measuring liquids. When you put water in a measuring cup and hold it at eye level to look at it, you will see that the surface of the water curves downward. The bottom of the curve in a column of water is called a **meniscus,** and you want to measure that, not the top edges. Also, if you use an eyedropper, be careful to let only one drop fall at a time.

MENISCUS ➞

Always measure everything you can. Whenever you describe anything, try to measure it. Note the length, weight, volume, area, temperature, age, number of things, and the time when things occur. When you are making one kind of measurement of each object in a group, use the same tool to measure each object, so that there is no difference in measurement due to differently marked tools, called **calibrated tools**. When measuring the length or volume of things that may expand when heated, measure them all at the same temperature, and record that temperature along with the measurement of length or volume.

KIDSOURCE TIPS

Here are some handy measuring tools you can find around the house:

TO MEASURE LENGTH

measuring tape

ruler

yardstick

meter stick

TO MEASURE WEIGHT

postage scale

bathroom scale

food scale

TO MEASURE VOLUME

measuring spoons

eyedropper

measuring cups

pint, quart, gallon milk
   cartons

liter soft drink bottles

TO MEASURE TEMPERATURE

body temperature
   thermometer

indoor/outdoor air
   temperature thermometer

candy or cooking
   thermometer

TO MEASURE AIR PRESSURE

barometer

tire pressure gauge

CHOOSING THE PERFECT PROJECT

"I like baseball. I love baseball. So I was going to find out what makes the perfect pitch. My project was great. It was perfect for me."

—Raphael Navarro, age 13

Now that you have the basics of the scientific method, you need an idea for your science project! What will it be? Let KidSource help you create an original, exciting, and interesting project!

HOW TO BE ORIGINAL

Scientists are explorers and discoverers who learn new things about our world by building on what they know. You want your project to find out something new, to add a building block to the knowledge about our world. Who knows? Your project could help scientists make more discoveries.

The best way to have an original science fair project—and to have the most fun—is to explore something that you care about.

GET SPECIFIC

The more specific your topic is, the more original your project is likely to be. If your topic is fishing and your question is

"What do fish like to eat?" get more specific. Choose a certain pond, lake, or stream and find out what kind of fish make it their home. Find out, too, what kind of food is available there for them. Perhaps trout live in a nearby stream where there are plenty of flies. Now you can ask, "Which flies do trout in Red Feather Lake like to eat?"

Say your topic is skiing and your question is "What kind of person makes the best skier?" Get more specific and choose one **characteristic** of skiing to study, such as speed or control, and one characteristic of skiers to compare, such as age, experience, muscle mass, or length of legs. For example, you could ask, "Do the fastest skiers have the longest legs?"

TRY FIELD WORK

Any field work you do is guaranteed to be original. That's because you are looking at one special place at one special time, something no one, including even you, can duplicate. If you study the insects in your backyard, the **fossils** on your hiking trail, the chemicals in your drinking water, the pollution in your river, or the traffic on your street, your work will be original and could turn out to be very significant to your community.

That's Personal!

Andrew Ritter, a 14-year-old from The Brentwood School in Los Angeles, California, came up with an original idea for a science project, using himself as the subject! Andrew had difficulty digesting dairy foods and was very motivated to find a solution to his problem.

So, Andrew did some research and discovered a theory controversial among scientists. It suggests that a person can adjust to lactose (found in milk) through small daily dosages of milk products. After talking to a researcher at Purdue University, and with his parents' approval, Andrew decided to experiment with this method on his own body.

Six weeks later, Andrew was eating cheese omelets, lasagna, and his first cheeseburger. Now, that's an original project!

(Note that you should never do a potentially harmful experiment on anyone, including yourself.)

TEST ADVERTISING CLAIMS

Check out advertising claims on products. With new products constantly being developed, you can find one to use for an original science fair project. Test to see if that new antibiotic soap really does kill bacteria, if a new cleaning product actually can remove any stain, or if a new plant food really does make tomatoes grow bigger. Or you could test for vitamin C content in a sports drink. You could also compare product effectiveness. Which dishwashing liquid cuts the most grease? Which contact lens solution is the most sterile?

DO RESEARCH

Research is looking into the collection of knowledge that exists about a subject. You can ensure that your project is original when you do your research. As you learn about your topic, you will discover whether or not others have already asked your question and if they found any answers. If your question is answered in your research, then you need to go back and think of another question. Or if you find that the procedure you are planning has already been done, you can change it by altering another variable. Your research will tell you what has been done and is now known, which is why it is so important to do research before you start your project.

You Can Contribute!

Seventh-grader Dara Weinberg won first place at the Los Angeles County Science Fair in 1996 for her project, "Effects of Charcoal Ash on Mealworm Growth and Development." Dara wanted to do something that could help people, something environmentally relevant. Her study showed some of the effects of air pollution on mealworms. "I felt that I liked science, but it was an unreachable thing. I didn't know that kids could do it. I didn't know that I could do it. That I'd done an experiment that would mean something to the world was the best feeling."

CREATING YOUR PERSONAL PROJECT

Here is your chance to create a science fair project personally exciting to you. Just follow the next few steps to come up with your original project, complete with a hypothesis and a procedure. If you already have a project idea, you can go on to Perfecting Your Project Idea.

CHOOSE A TOPIC THAT FITS YOU

Think of a few activities that you enjoy doing in your free time. Choose one. Whatever it is, it is important to you, and you know something about it.

FOLLOW THE QUESTIONS

Write down what your activity is. This will head your **list A** (see sample on page 33). Next, think about your activity. What happens when you do this activity? Do you do this activity inside or outside? Do you use equipment? Do you do this activity with other people? When do you do it? Do you need special skills or attributes or practice to do it? Why do you think you are interested in it?

Now you can add these things to list A:

WHERE you pursue your interest. What things describe or affect the place, or environment, where you do your activity? If you do it outdoors, you can list the climate, the temperature, the **terrain** (or ground condition), the currents and tides (of rivers, lakes, and oceans), and all its man-made features, such as buildings, fields, gardens, fences, and streets.

If your interest involves something you do indoors, describe where in particular you do it, including the size and shape of the space, the temperature and humidity there, how the area looks and sounds, and any other important features.

WHEN you pursue your interest. What time of the day or night, what part of the week or month, and which season is best for doing your activity?

WHO else may be involved. Does your activity involve other people? How about plants or animals?

WHAT you need. What steps do you follow to do your activity? You probably need some special skills or abilities to do your activity. What type of uniform or clothes do you wear? What tools, supplies, equipment, or machinery do you use?

SAMPLE ACTIVITIES

football	fishing	swimming	chess
skating	gardening	biking	television
movies	computer games	reading	card playing
baseball	cooking	sewing	painting
hiking	puzzles	shopping	flying kites

SAMPLE LIST A
Gardening

Where	When	Who	What
outdoors	days	insects	flowers
water	summer	birds	gloves
rocks	spring	butterflies	vegetables
sand	rain	slugs	weeding
soil	wind	rabbits	hoes
	sunshine		roots
			pollination
			cultivation

Now choose one thing from your list A, and start **list B** (see sample on page 35). Write at least five questions about that thing, using these questions as guides:

Why?

Why is it there?

Why does it do what it does?

Why do you need it?

Why is it the way it is?

When?

When does it happen?

When does it do what it does?

When does it behave or respond one way?

When does it behave or respond differently?

When did it become the way it is?

When did it look or seem different?

When is it easy to use?

When is it a problem?

When is it easy to notice?

Where?

Where does it happen?

Where does it come from?

Where does it go?

Where does it have its greatest effect?

Where is it hard to do or see or hear or feel?

Where is it best?

How?

How many are there?

How does it work?

How did it get there?

How does it happen?

How did you notice it?

How did others notice it?

How did it affect something?

How did others affect it?

How did it sound?

How did it look?

How did it move?

How did it smell?

What?

What is it doing?

What is happening to it?

What is it for?

What is special about it?

What is causing it to happen?

What would be different if one thing about it changed?

SAMPLE LIST B
Pollination

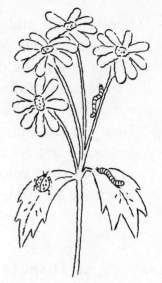

1. Why is pollen yellow?

2. When are plants pollinated?

3. Where does the pollen go?

4. How do insects affect pollination?

5. What would happen if plants were not pollinated?

From list B, pick the question that seems the most interesting to you to be your project topic. Write this question on the top of a clean sheet of paper to start **list C**.

Consider this question and think of all the things you **DO** know about your topic. Go to your library and research your topic. Then, on list C, make a list of your guesses of possible answers to your question, based on what you already know. These are your possible hypotheses.

SAMPLE LIST C

How Do Insects Affect Pollination?

1. Insects carry pollen to flowers.

2. Insects choose flowers that are certain colors.

3. Insects collect pollen.

4. Insects brush pollen on other insects.

Look at list C and pick the hypothesis you think is most likely to be right. Write down your hypothesis to start **list D**. Next, think of at least two or three things you could observe and measure that might test for your answer, and write those down on list D. Then describe how you would go about each test and what you think you would need—that is your procedure.

Now you are ready to show your list D to your parents or guardian and your teacher and ask them if they think one of these ideas would make a good project. Based on your time and available supplies, choose the best one to be your science fair project.

SAMPLE LIST D

Insects Carry Pollen to Pollinate Flowers

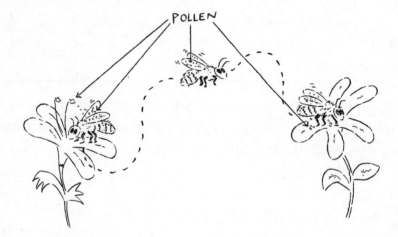

1. I could watch the insects in my garden to see how they affect the tomato plants. I would need a garden, tomato seeds, a watering can, a chair, time to watch the garden very carefully, and maybe a magnifying lens.

2. I could capture insects before and after they visit flowers and examine them. I would need a garden with insects and flowers, a net and a capture jar, a magnifying lens, and maybe a microscope.

3. I could examine flowers before and after insects have visited them. I would need a garden with flowering plants and a magnifying glass.

4. I could cover some flowering plants in my garden with a tight mesh gauze so that no insects can reach them, and then compare them to similar flowering plants that insects can get to. I would need a garden with flowering plants, gauze and tape, and a magnifying lens.

PERFECTING YOUR PROJECT IDEA

Now you have a rough sketch of your project outline, with the first draft of the four basic parts. You have your question, or project topic, which describes what you want to find out, or accomplish, in your project. You have your best guess, or hypothesis, which predicts what you think will happen. You have your procedure, the steps of your experiment or field work. And you have a list of the **materials** you will need to do your project. Let's look at these components, one by one, and fine tune them.

THE QUESTION

First of all, can your question be answered best with an experiment or field work? If you need to observe things **as they are** in the world, then your project will be field work. Make sure your question is specific—that you are looking at one thing during one time in one place. If your question is "What do insects eat?" narrow it down to a specific insect in a specific place at a specific

LET'S BE REAL!

You may be wondering if you can blow bubbles in outer space or what the ocean floor looks like or how a baby elephant grows into an adult. Do you really have the equipment, the skills, or the time to find these things out? Do you have a space ship or an ocean liner, do you have space flight training or know how to scuba dive, and do you have years to watch an elephant grow up before your project is due? Chances are, you don't. Although it is probably not practical to explore these kinds of questions in this science fair project, save them anyway—there may be another time and another way to look for answers.

time. For example, you could ask, "What do grasshoppers eat in the summer?"

If you need to observe things in a controlled test, then your project will be an experiment. Make sure you have narrowed your question down so you can test for something specific. If your question is "What happens to fruit juice when it is left out uncovered?" choose an exact amount of one kind of juice left out uncovered for a specific time, and zero in on one characteristic to test for, such as vitamin C content.

To help you narrow down your question, here are some suggestions:

• Narrow down to two comparisons.

For example, compare the fingerprints of members of your immediate family with the fingerprints of the immediate family members of a friend.

• Narrow down to specific time periods.

For example, for television studies, use specific programming, like news hours, sitcoms, hour dramas, or Saturday morning cartoon shows.

• Narrow down to very small-scale environments.

For example, to study small plants or animals, choose to look at a single plant or rock or an area of about a square foot.

THE HYPOTHESIS

Most good hypotheses are easy to spot. They usually take the form of an "if . . . then" statement and so have two parts: first a description of what is changing and then a description of what you expect to happen.

If You Are Doing Field Work

Your hypothesis states when and where you will go to make your observation or gather your data, as well as what you expect to see or collect there at that time. For instance, if you wanted to find out which birds nest in oak trees, your hypothesis would state:

> *If I go to the state park in March and watch an oak tree, then I will be able to study the birds nesting there.*

If you saw a pretty garden spider in your yard and found out what kind it was, you might come up with a question like "How often does an *Argiope aurantia* eat?" When you did your research, you found out that this spider likes sunny locations and rebuilds its big web each night. Based on what you know, your hypothesis could state:

> *Because the **Argiope aurantia** rebuilds its web each night, I think it eats every day. If I look at the spider's web in my backyard each morning for two weeks, then I can observe what it catches and when it eats.*

When You Do an Experiment

Your hypothesis should state what you predict will happen when you change one thing in your experimental group. It should describe your control and experimental groups, as well as the changing variable and your prediction—the responding variable.

For instance, if you are testing whether or not your neighbors' pulse rates change after they pet their dogs, your hypothesis would state:

> *If my neighbors pet their dogs, then their pulse rates will go down.*

The groups in this experiment are the neighbors before they pet their dogs (the control) and the neighbors after they pet their dogs (experimental). The changing variable is the action of petting the dogs, and the responding variable is pulse rate.

Model Mistakes

A model is an illustration of something that is already known to scientists. Models are interesting, and they are great for learning about scientific ideas, but they are not original, and they are not suitable as science fair projects. Also avoid using any prepackaged science kits in your project, unless you uniquely alter them in some way to "do" something other than what they were designed to do.

THE PROCEDURE

The procedure is made up of two parts: the materials and the instructions.

If you are doing field work, your procedure would list all the tools used and explain how you collect, measure, and describe your observations. For example, if you are observing a **tide pool** every day at low tide, your procedure might look like this:

Materials

- ruler
- thermometer
- pencil and paper
- camera and film
- map
- watch
- magnifying glass
- notebook
- pen

Directions

1. Record the time, date, place, temperature, weather, and water level.

2. Measure the area of the space you are observing.

3. Draw or photograph the animals you see.

4. Count the different kinds of animals.

5. Count the individuals of each kind of animal.

6. Measure the size of each individual.

7. Describe any activity you observe.

8. Repeat steps 1 through 7 for two weeks.

If you are doing an experiment, your procedure would describe how you arrange, collect, or build your control and experimental groups, how you measure and apply your changing variable, and how you measure your responding variable.

Your procedure would also list how many subjects you used and how many trials you did. For example, let's say your hypothesis states:

> *If a pendulum has a longer string, then it will swing in slower arcs.*

Your procedure might look like this:

KIDSOURCE TIPS

Make sure you are not planning anything specific to a season if you won't be experiencing that season during your project.

Materials

- ruler
- scissors
- string
- five fishing leads of equal weight
- stopwatch

Directions

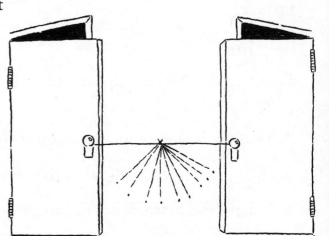

1. Measure and cut five different lengths of string: 15 cm, 30 cm, 50 cm, 60 cm, and 75 cm.

2. Attach one fishing lead to each string to make a variety of pendulums.

3. Tie a meter-long string straight across the backs of two chairs or to two doorknobs in a room, making sure this string is high enough for the longest pendulum string to swing free on it.

4. Attach one pendulum to the string, hold the pendulum 90 degrees from the floor, and release it to make it swing.

5. Measure how far the pendulum swings on either side of dead center. Record your results.

6. Measure how long it takes the pendulum to swing from one side to the other. Record your results.

7. Repeat steps 3 through 5 for each pendulum, and compare the results.

SAFETY

Safety is a consideration of the utmost importance when you are planning a science project. Follow these tips to protect yourself and others:

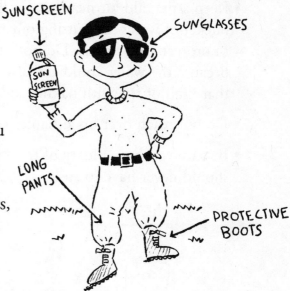

- Always have a teacher, parent, or other responsible adult supervise you, unless after you have explained exactly what you are going to do, the adult gives you permission to work on your own.

- Pay attention and be alert while you are doing your experiment. Fooling around can be very dangerous.

- If you are doing field work outdoors, dress appropriately, with protective boots, long pants, sunglasses, and sunscreen, if needed. Avoid any poisonous plants and threatening animals.

- Wear protective goggles, thick gloves, and a lab apron whenever you work with dangerous materials or chemicals. Remove or tie up any loose clothing or hair. Work on a protected surface and organize your materials so that they are within your reach. Put away anything you do not need, and keep your work area neat.

- If you are using any dangerous materials or chemicals, read all the labels before beginning your experiment.

- Do not inhale fumes produced in any chemical reactions. Do not taste chemicals or solutions.

- Do not eat or drink anything while you are doing your experiment. Clean your hands carefully after you finish working.

- Watch out for sharp edges, and be careful handling glass objects. Do not touch glass that has been recently heated.

- Never tinker with household electricity or power or gas lines. Do not touch any high-voltage source or anything connected to one. Never put an electrical device in water.

- Keep a first aid kit nearby, as well as a fire extinguisher and a bucket of water for putting out possible fires. Have hand mitts ready to protect your hands. Do not put water on an oil-based fire, because the water could actually spread the fire. Instead, smother that kind of fire with baking soda or a blanket.

- Do not reach over open flames or heating instruments.

- Be sure you have plenty of time to run your experiment safely. You should never have to rush.

THE INFORMATION DIRECTOR

"When I was a child, I called up an expert at a museum to ask what I thought was an important question. I was treated so kindly, that may have started me on my path in science."

—Dr. E. C. Krupp, Griffith Observatory, Los Angeles, California

To do an original science project, you need to know what has already been done on the topic you are exploring. And you need to base your hypothesis on facts you know about your topic. You also need to know how to find experts to help you and where to find materials you need.

The Information Director will help you find all you need to know. First, the Information Director takes you through libraries, helping you with books and periodicals. Then the Information Director helps you contact experts and find all the supplies you need to do your science fair project. Let the Information Director help you do the best science fair project ever!

LIBRARIES

Libraries are an amazing source of information and also a lot of fun to explore. Plus their services are all free—or very close to it! You can find out almost anything at a library, and usually, if some information you need is not there, a librarian can order it for you. University and college libraries and your city's central public library

KIDSOURCE TIPS

Here's your library checklist. Don't forget
- a bag for carrying books home
- your library card
- change for photocopies
- pens and notecards.

have the most material available. Smaller local public libraries sometimes specialize in one topic and generally have less material.

You will need a library card to check out books and videotapes, but reference books, magazines, newspapers, and microfiche (also called **microfilm**) must be examined at the library. Be sure to bring pens and notecards to write down your sources of information and to take notes from the materials you look at in the library. You will also want to bring change to make photocopies. Here are a few good places to start your research:

ENCYCLOPEDIAS

- First look at all available encyclopedias to get basic information on your subject. Different encyclopedias will have different information on a subject. Make photocopies of the necessary pages or take notes, keeping track of which encyclopedias you're using. This background will help you understand the more detailed information you will find in books and magazine articles.

COMPUTER SYSTEM/CARD CATALOG

- Next, try the computer system search or card catalog to look for books on your topic. Use the key words from your topic question to search under "subjects." Write down the titles, authors, and call numbers—the numbers that tell you where in the library the book can be found. Also write down the copyright date. This date is important because it tells you when the book was published, and the older the book is, the less up-to-date the information in it may be. If you are researching new technologies or recent discoveries, read the newest books you can find. Look in both the juvenile book collection and the adult collection. Try to find two or three books that you can read on your topic.

Write a brief summary of what you read in each book, with any direct quotes you think are important. On the same page as your summary and notes, write down the author, title, publisher, date, and city where the book was published. If you use any of the book's information in your project, give the author credit. Also, it's important to keep track of everything you have read on your topic to include in your **bibliography,** which is a list of research sources. (See more on your bibliography on page 48.)

KIDSOURCE TIPS

Always know where the nearest dictionary is! You will want to look up new words as you come across them.

PERIODICALS

- When you have a solid understanding of your subject from books, you are ready to look at the more recent publications, magazines, and newspapers, also called **periodicals**.

- One main guide to periodicals is called the *Readers' Guide to Periodical Literature.* This lists all the articles in magazines and journals on your topic. Look up your topic in the *Reader's Guide.* Once you have found an article that sounds like it may have information you need, write down the name of the article, its author, the magazine or journal it was published in, and the volume, number, date, and page number indicated.

- To find out when important events concerning your topic happened, look in an index called *Facts on File.* You can look up your subject, using the same key words you used with the card catalog, and find out when articles concerning it were published. Microfilm is small film that stores the pictures of the pages of the periodical for projection on a screen. You can read the article and take notes, making sure to include bibliographic information. If the article is very helpful, you may want to photocopy it, using a machine the library will have. The librarian can show you how this works.

- You also will want to look at newspaper articles, another source of recent information. Some newspapers let you find articles through the Internet, and your library may have Internet access. By going to your central public library, you can have access to more than one newspaper and to many that are not local.

Your librarian can tell you how to look up articles by subject and date in the newspapers stored at your library. If your library has CD-ROM, you may be using a system called *Infotrac,* or you may use a system called *Newsbank CD-News.* Again, whether you make copies of articles or take notes, you need to write down bibliographical information for each source.

YOUR BIBLIOGRAPHY

While you are doing your research, you should make your bibliography. The bibliography will appear at the end of your project report. It should list all the materials you read and should also note, under "other sources," all the experts you talked to, what they talked about, who they are, and where they are from. The judges will know how thoroughly you did your research when they see your bibliography.

A bibliography should have all the information anyone would need to find the sources you used. Alphabetize your list by author's last name, and when there is more than one author for one book or article, list the authors alphabetically. Here are some formats to follow:

For Books:

Last Name of Author, A. A. Title of Book. Location of publisher: Name of Publisher, copyright date.

EXAMPLE: Weaver, I. M. *Spiders and Their Webs.* New York: Great Books, Inc., 1993.

For Periodicals:

Last Name of Author, A. A., Last Name of Author, B. B., & Last Name of Author, C. C. (year#, Month day# of publication). Title of Article. Title of Periodical, Volume #, page #–page #.

EXAMPLE: Potato, C. A., Watcher, T. V., & Zombie, B. A. (1996, June 10). "Commercials and Snacking Habits." *Programming Guide,* vol. 7, pp. 34–37.

For Newspapers:

Last Name of Author, A. A. (year#, Month day# of publication). Title of Article. Title of Newspaper, pp. A1, A4.

EXAMPLE: Beans, F. O. (1996, April 1). "People Don't Read." *Daily Rag,* pp. A1, B23–24.

Along with the page or pages, be sure to list the section of the newspaper where you found an article—for example, an article on the front page would be A1. Magazines may include an issue number, which comes after the volume number, and you should add this in parentheses.

TALKING WITH EXPERTS

Government agencies, universities, museums, and science organizations where you can contact scientists and other people with expertise are an excellent resource for your project. They can be especially helpful if you get stuck on something during your research or procedure that you don't understand. Also, if you are doing field work, some experts may be interested in your

data or collection. You may want to ask experts for tips on collecting and preserving specimens or on experimental techniques unique to your topic. You may want to clear up questions from an article that you read. Sometimes science groups and even businesses will let you use their equipment or help you operate equipment that is complicated. You also may find you have contacted someone who can be a **mentor** to you—some-one who can help you develop as a scientist!

EXPERTS ARE FOR EXTRA HELP

When talking to professionals, you should look for special information that will help and guide you, not general information that you could find in a library. Never expect an expert to tell you exactly how to plan your science fair project or how it should turn out. Experts appreciate your enthusiasm and your interest—that is what motivates them to help you!

BE PREPARED

Before you try to contact an expert or professional, make a short list of questions you will ask. Make your questions "open-ended," which means the answers to your questions will be more than just a yes or no. Also, because in most cases you are talking to busy professionals, make your questions very specific.

Show your questions to your science teacher and ask if they are appropriate and clear. Do you need to do more of your own research? Also ask your teacher if you have picked the best expert to question. While a marine biologist may know about comets, and an astronomer may know about sharks, it is better to ask a marine biologist about sharks and an astronomer about comets.

While you are at the library, ask your librarian for tips on getting experts' phone numbers and addresses. Libraries often have phone books for many cities, as well as many professional and business indexes with names, addresses, and phone numbers.

When you call someplace, looking for an expert to talk to, here are some hints:

- Have ready your list of questions and paper and a pen to take notes. When you get someone on the line, say hello and give your name. Ask for the address of the location in case you need to visit it for an interview. If the mailing address is different, you'll need to get that, too, in case you need to send the expert a list of questions or when sending a thank-you note.

- Explain that you are doing research for a science fair project and would like to ask an expert a few questions. Ask if there is someone available you may speak to.

- If the person on the line says yes, ask for the spelling of the expert's name, and be sure you can pronounce it. Then thank the person for helping you, and either ask to be connected or request the expert's phone number or extension.

- When your expert answers, give your name and say that you are researching your topic for a science fair project and would like to ask a few questions.

- If your expert agrees to answer some questions, ask when would be a good time. Be ready to ask your questions then and there, or to make an appointment to call or visit at a later time. If you set up a meeting in the future, be sure you make it at a time a parent or guardian can bring you.

KIDSOURCE TIPS

When you are arranging to call or meet with an expert, know for sure when you will be available for an appointment before you agree to a time.

Always be on time for your appointment, and call as soon as you can if you must cancel or postpone an appointment.

Don't forget to thank people for taking the time to speak with you.

• Cross off the questions you have asked, and identify which question is being answered in your notes. Don't take more than fifteen minutes of your expert's time. If you want to tape-record your conversation to keep from missing anything, be sure to first ask the expert for permission.

• When you are done, thank the expert for her time before you say good-bye. Immediately after your interview, write the expert a thank-you note.

• After the science fair, drop your expert another note telling her how you used her information and how your project went.

KIDSOURCE TIPS

Be sure to allow enough time to receive an answer if you write for information. A month should be appropriate.

When you first try to contact an expert, ask if that person has the time to help you.

Be prepared to ask your questions the first time you speak with the expert—he or she may actually have a free moment when you first call.

A SAMPLE CONVERSATION

Suppose you were interested in butterflies from your area. A phone call to your local natural history museum might go like this:

YOU   Hello. I would like to talk to someone who is an expert on butterflies. Is there anyone available I could speak with?

THEY   Yes, Dr. Blackwing is our head entomologist. Let me see if he is available right now.

YOU   Thank you. Could you please spell Dr. Blackwing's name for me?

THEY   Sure. It's B-L-A-C-K-W-I-N-G. One moment, please. (There is a pause while you are put on hold.)

THEY   Hello? I'll put you through to Dr. Blackwing now.

YOU   Thank you.

DR. BLACKWING   Hello. I'm Dr. Blackwing.

YOU   Hi, Dr. Blackwing. Thank you for talking with me. My name is (your name), and I am doing a science fair project on local butterflies. I was wondering if I could ask you some questions about butterflies when you have the time.

DR. BLACKWING — I actually have about fifteen minutes right now. Do you have your questions ready?

YOU — Yes, I do. First, I was wondering if there were butterfly collections at the museum that would help me identify some butterflies I have collected.

DR. BLACKWING — Yes, we do have a collection of local butterflies. You may call our archivist, Ms. Swallowtail, to make an appointment to look at them.

YOU — Great! How can I reach her?

DR. BLACKWING — Her number is 555-1234. She's usually in every afternoon from 1:00 P.M. to 5:00 P.M.

YOU — Okay, thank you. Can you tell me how you identify different **species** of butterflies?

DR. BLACKWING — Certainly, although it would be easier to understand if I could explain it while you were actually looking at the different butterflies in the collection.

YOU — Could I make an appointment to come in and meet with you?

KIDSOURCE TIPS

If you are calling someone long distance, ask your parents or guardian for permission before you use the phone.

Sometimes it takes effort to contact an expert. You may have to try many different sources before you have any success. Most of the time, politeness and persistence will pay off.

DR. BLACKWING   Sure, Tuesday afternoons at 4:00 P.M. are a good time for me.

YOU   I think my mom can take me to the museum then. May I check with her and call you back this afternoon to confirm?

DR. BLACKWING   Certainly.

YOU   May I bring my collection with me?

DR. BLACKWING   Absolutely. You may have specimens of species we don't have. I'd be very interested in seeing what you have found.

YOU   Would it be all right if I ask the rest of my questions on Tuesday afternoon?

DR. BLACKWING   Absolutely.

YOU   Great! Thank you very much, Dr. Blackwing. I appreciate your time.

DR. BLACKWING   You're welcome. Good-bye.

YOU   Good-bye.

SCIENCE SOURCES

Here are descriptions of different science categories you will find at most science fairs, along with local and national organizations that can tell you more about a particular study of science. Selected books and magazines in each area that may be helpful to you are also listed. Plus you'll find details on several science fair projects to give you an idea of different things you can do.

Astronomy: the study of the Sun, planets, stars, and other objects in outer space and their composition, motion, size, distance, and other characteristics.

Sources

Local

Look in your phone book for **planetariums, observatories,** and universities near you. Check your newspaper for current shows or exhibits at these places. Your newspaper will also have science articles and listings of lectures and astronomy club meetings and "star parties"—gatherings where local astronomers bring telescopes to view interesting objects in the sky.

National

National Aeronautics and Space Administration
300 E St. SW
Washington, DC 20546
(202) 358-0000 ☛ Web site: **http://www.NASA.gov**

Books

Gardner, R. *Projects in Space Science.* New York: Julian Messner, 1988.

Greenleaf, P. *Experiments in Space Science.* New York: Arco Publishing, 1981.

Moore, P. *The Amateur Astronomer.* New York: W. W. Norton & Company, 1990.

> ### Science Fair Project
> ## Collecting Micrometeorites

In this project, you go out into the field to get samples and bring them home to analyze them.

 Background:

Pieces of objects from space, called **micrometeorites,** continuously fall to the earth's surface. Annual meteor showers increase the number of particles that fall to the ground. Many meteorites have a high iron content. How can I collect and identify micrometeorites?

Annual Meteor Showers	Date
Bootids	January 2–3
Lyrids	April 19–22
Perseids	August 9–14
Orionids	October 16–20
Leonids	November 14–18
Andromedids	November 17–23
Geminids	December 9–12

* *This list is approximate. Check your local newspaper for specific times in your area.*

Hypothesis:

If I collect atmospheric particles during, before, and after a meteor shower, I should then collect the most micrometeorites during the meteor showers.

Materials:

- clean glass pie pan
- magnet
- plastic bag
- permanent marking pen
- sewing needle
- microscope slide
- microscope or magnifying glass

Procedure:

1. Check the chart above to find when an annual meteor shower occurs, and pick one evening during a shower to place a clean glass pie pan outside overnight.

2. The next morning, cover a magnet with a plastic bag. Sweep the covered magnet over the bottom of the pie pan to collect metallic particles.

3. Gently turn the plastic bag inside out over the pan, keeping all the metallic particles inside, and seal it. Use a permanent marker to write the date on the bag.

a

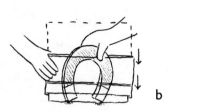

b

c

d

4. Repeat steps 1 through 3 at least three times during meteor showers.

5. Repeat steps 1 through 3 at least three times several weeks before and after a meteor shower.

6. Magnetize a sewing needle by rubbing it in one direction on the magnet. Set it aside.

7. Open one plastic bag and pick up one particle at a time with the magnetized needle and place it on a microscope slide, sorted by size.

8. Examine the particles under a microscope or with a magnifying glass. Write down descriptions of how the particles look, including shape, relative size, shininess, and number of particles similar to one another. Record the date of collection with these data.

9. Repeat steps 6 through 8 for each bag of particles.

10. Compare the data from the samples.

Biology: the study of the origin, development, structure, diversity, and behavior of all living things, including
Microbiology—the study of living things too small to be seen with the naked eye.
Botany—the study of the structure and growth of plants.
Zoology—the study of animals in their natural habitat.

Sources

Local

Look in your phone book for local **arboretums,** gardens, zoos, animal reserves, aquariums, fish and pet stores, **marinas,** natural history museums, and universities. Check your newspaper for science articles and lectures, demonstrations, and workshop opportunities at parks, marinas, zoos, and museums.

National

U.S. Department of Agriculture
Agricultural Research Service
Room 340A Administration Building
Washington, DC 20250

U.S. Forest Service
PO Box 96090
Washington, DC 20090
(202) 205-1248

U.S. Fish and Wildlife Service
Main Interior Building
1849 C St. NW
Washington, DC 20240
Megan Durham (202) 208-4131 ☛ Web site: **http://www.fws.gov**

National Park Service
U.S. Department of the Interior
Washington, DC 20240

The Sierra Club
85 2nd St., 2nd fl.
San Francisco, CA 94105
(415) 977-5500 ☛ Web site: **http://www.sierraclub.org**

The National Audubon Society
700 Broadway
New York, NY 10003
(212) 979-3000 ☛ Web site: **http://www.audubon.org**

Books and Magazines

Hershey, D. R. *Plant Biology Science Projects.* New York: John Wiley & Sons, 1995.

Cain, N. W. *Animal Behavior Science Projects.* New York: John Wiley & Sons, 1995.

Ranger Rick
Membership Services
National Wildlife Federation
8925 Leesburg Pike
Vienna, VA 22180-0001

Zoobooks
3590 Kettner Blvd.
San Diego, CA 92101

Science Fair Project
Strange Seeds in My Backyard

In this experiment, your control is all the plants you see growing in your backyard, plants you could reasonably expect to deposit seeds into your soil samples. Your changing variable is nature's many means of seed transport—the wind, rain, and animals that carry exotic seeds into your backyard. The responding variable is the appearance of unknown plants in your experimental gardens.

Background:

Seeds are dispersed in many ways: carried by wind and rain and in animals' fur, gathered and carried along by animals for food, and eaten and redeposited by birds. Can I find exotic seeds in my own backyard and find out how far they traveled?

Hypothesis:

If I dig up soil samples in my own backyard, I can then grow seeds that are present in the soil. And when those plants are grown, I can then identify them and look for their possible parent plants outside my backyard.

Materials:
- water
- chlorine bleach
- five clay pots
- five glass jars that fit over the pots
- small garden spade or trowel
- garden, horticulture, and wildflower books
- bicycle
- pencil and drawing paper or camera and film

Procedure:

1. Take a plant census of your
backyard, counting and
describing all the plants you
can find growing there.
Find out the names of
the plants in a garden
or horticultural book.
To identify some of the
more difficult ones, it may
be helpful to contact a botanist.

2. Mix 2 liters of water and 15 milliliters of chlorine bleach. With this
mixture, clean and sterilize five clay pots and five clear glass jars that
fit over the pots. Allow these to dry.

3. In five different places in your
backyard, dig samples of
soil, and fill each pot with
one sample. Add enough
water to moisten, but not
soak, the soil, and then cover
each pot with a glass jar.

4. Put the pots in a sunny place in your
backyard, and check them daily for a
period of four to six weeks. Record
your descriptions and drawings or
photographs of any plants you observe
growing inside the pots. Each day, lift
the glass jar, but not above the pot rim, to
let in fresh air. Water when the soil begins to
dry out.

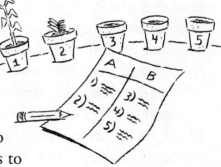

5. When the plants in the pots have grown enough that you can identify them in garden, horticultural, and wildflower books, list their names. Now make two lists: list A of the plants in the pots that are already growing in your backyard, and list B of the plants in the pots that are not currently growing in your backyard.

6. Look in your neighborhood for the possible parents of the plants on list B. First look next door, then walk down your street, then ride your bike through your neighborhood. If you still haven't found similar plants, ask your parents to drive you a few miles around your neighborhood to look for the possible parent plants.

Chemistry: the study of the composition and structure of substances.

Sources

Local

Look in your phone book for museums of science and industry, pharmaceutical companies, laboratories, universities, and research and development centers.

Check your newspaper for lectures, demonstrations, and workshops at museums and universities.

Check with your local Environmental Protection Agency office or Sierra Club—they may be interested in pollution data you gather.

National

American Chemical Society
1155 16th St. NW
Washington, DC 20036
(202) 872-4600 ☛ Web site: **http://www.acs.org**

The Environmental Protection Agency (EPA)
401 M St. SW
Washington, DC 20460
(202) 260-2090 ☛ Web site: **http://www.epa.gov**

The Sierra Club
85 2nd St., 2nd fl.
San Francisco, CA 94105
(415) 977-5500 ☛ Web site: **http://www.sierraclub.org**

Books

Science-in-Action: The Marshall Cavendish Guide to Projects and Experiments Fun With Chemistry. New York: Marshall Cavendish, 1989.

More Science Experiments You Can Eat. New York: J. B. Lippincott, 1979.

> ### *Science Fair Project*
> ## Freezing Rates of Pure and Unpure Water

This experiment compares pure water (the control) with seven changing variables to discover if these variables affect the responding variable—the time it takes water to freeze solid.

Background:

Water boils faster when salt is added to it. Will salt and other added substances affect how quickly water freezes?

Hypothesis:

If substances are added to water, then the mixtures will freeze more slowly than pure water alone.

Materials:

- three plastic ice cube trays
- marker
- seven bowls
- distilled water
- isopropyl alcohol, granulated sugar, corn oil, salt, vinegar, baking soda, gelatin
- seven spoons
- graduated cylinders

Procedure:

1. Take three clean plastic ice cube trays. Label one tray "pure water." Label each of the other two trays like this: On the front of the first row (of two cubes), write "alcohol"; on the second, "sugar"; on the third, "corn oil"; on the fourth, "salt"; on the fifth, "vinegar"; on the sixth, "baking soda"; and on the seventh, "gelatin."

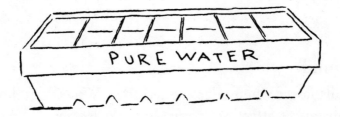

2. Fill seven small mixing bowls with ¼ liter of distilled water. Add 15 milliliters of alcohol to the first bowl, 15 milliliters of sugar to the second bowl, 15 milliliters of corn oil to the third bowl, 15 milliliters of salt to the fourth bowl, 15 milliliters of vinegar to the fifth bowl, 15 milliliters of baking soda to the sixth bowl, and 15 milliliters of gelatin to the seventh bowl. Use a separate spoon to stir each bowl until all or most of the substance in the water has dissolved.

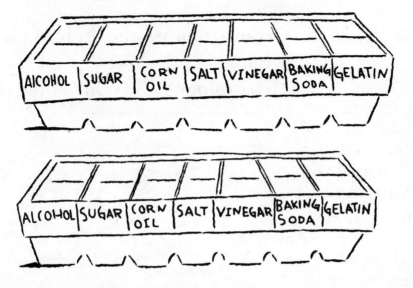

3. Take the two trays that are labeled with the seven substances and fill each of all four cubes labeled "alcohol" with one-fourth of the alcohol mixture. Fill each of the other four sets of cubes with one-fourth of the mixture indicated by the label. Fill all the third tray's cubes with pure distilled water.

4. Place all three trays in a freezer.

5. For one day, check the trays every fifteen minutes for four hours. At each quarter-hour interval, record which cubes are frozen and describe how each cube looks as it starts to freeze.

Earth Sciences: the study of the earth, including

Geology—the study of the composition and physical characteristics of the earth.

Paleontology—the study of fossil remains of living things from prehistoric times.

Oceanography—the study of the world's oceans.

Meteorology—the study of the weather.

Sources

Local

Look in your phone book for natural history museums, rock shops, weather stations, and universities. Check your newspaper for rock and mineral club meetings, gem shows, fossil hunts, and paleontological societies. Call your park service to see if there are fossil sites or weather stations near you.

National

The U.S. Geological Survey
12201 Sunrise Valley Dr.
Mailstop 119
Reston, VA 20192
(703) 648-4000 ☛ Web site: **http://www.usgs.gov**

The U.S. Bureau of Land Management
Resource Use and Protection
Main Interior Building
Washington, DC 20240

The National Weather Service Forecast Office
1325 East-West Hwy.
Silver Spring, MD 20910
(301) 713-0622 ☛ Web site: **http://www.nws.noaa.gov**

American Meteorological Society (AMS)
45 Beacon St.
Boston, MA 02108
(617) 227-2425 ☛ Web site: **http://www.ametsoc.org/ams**

National Oceanic and Atmospheric Administration (NOAA)
U.S. Department of Commerce
Washington, DC 20230
(301) 713-4000 ☛ Web site: **http://www.noaa.gov**

Books

Dixon, D. *The Practical Geologist.* New York: Simon & Schuster, 1992.

Reifsnyder, William E. *Weathering the Wilderness.* San Francisco: Sierra Club Books, 1980.

Sloane, E. *The Book of Storms.* New York: Duell, Sloan and Pearce, 1956.

> ## *Science Fair Project*
> ## Roadside Geology

In this example of field work, you go out into the field to record what you observe and also collect samples of rocks and fossils to study.

Background:

Highways are often made by blasting through hills of rock, which exposes layers of material that were deposited and then deformed millions of years ago. What can I discover at such a road cut?

Hypothesis:

If I go to a road cut, I can then examine exposed layers of rocks and try to identify the kinds of rocks and fossils I find.

Materials:

- notebook, pen, and pencil
- geological hammer
- plastic bags with marker
- geological map (of area with road cut), available at a bookstore or library or ordered from the United States Geological Service.

Procedure:

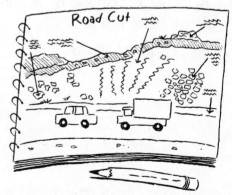

1. Go to a road cut, and in your notebook, describe where it is located. Draw a picture of the road cut. Describe the different layers you see. Describe any fossils you can see in each layer.

2. Use your hammer to remove a piece of rock from each layer. Describe each piece according to the following characteristics:

- the cleavage of the rock—the shape of it as it breaks away

- its streak—the color of the rock powder you get when you scrape it

- the rock's color and shininess

- the rock's hardness

Put each rock in a plastic bag and label the bag, noting which layer the rock came from.

3. Draw a picture of any loose fossils. Put them in a plastic bag and label the bag. Trace any flat fossils in the road cut, and indicate which layer each fossil was in.

4. Be sure your labeled specimens correspond clearly with your notes on them.

5. Get a geological map of the area, which identifies different rock types by different colors, and try to determine the names of the layers of rock you saw by matching your description of the rocks and fossils you examined with descriptions of the different rock layers on your map.

Physics: the study of the physical properties of things, such as force, motion, energy, heat, light, magnetism, sound, and electricity.

Sources

Local

Look in your phone book for museums of science and industry, observatories, planetariums, optics companies, engineering companies, civil engineers, special regional physics labs, and universities.

National

U.S. Naval Observatory
34th St. and Massachusetts Ave. NW
Washington, DC 20392
(202) 762-1467 ☛ Web site: **http://www.usno.navy.mil**

National Museum of Natural History
Smithsonian Institution
Washington, DC 20560
(202) 357-2700 ☛ Web site: **http://www.si.edu**

American Museum of Natural History/Hayden Planetarium
Central Park West at West 79th St.
New York, NY 10024
(212) 769-5100 ☛ Web site: **http://www.amnh.org**

Jet Propulsion Laboratory
4800 Oak Grove Dr.
Pasadena, CA 91109
(818) 354-4321 ☛ Web site: **http://www.jpl.nasa.gov**

Griffith Observatory
2800 E. Observatory Rd.
Los Angeles, CA 90027
(213) 664-1191 ☛ Web site: **http://www.griffithobs.org**

Optical Society of America
1816 Jefferson Pl. NW
Washington, DC 20036
(202) 223-8130 ☛ Web site: **http://www.osa.org**

Books

Science-in-Action: The Marshall Cavendish Guide to Projects and Experiments in Physics. New York: Marshall Cavendish, 1989.

Science-in-Action: The Marshall Cavendish Guide to Projects and Experiments in Light and Sound. New York: Marshall Cavendish, 1989.

Science Fair Project
Cool Batteries

This experiment tests for a responding variable, battery life, as affected by the changing variable, storage temperature, and as measured by tracing the circles of light the batteries produce in flashlights.

Background:

Batteries supposedly last longer when kept cool. How does temperature affect batteries?

Hypothesis:

If I compare batteries stored at different temperatures, I will then find that batteries stored at cooler temperatures last longer.

Materials:

- two 8-count packages of AA batteries, of different brands
- eight penlights that use two AA batteries
- four zipper-lock bags
- newsprint, tape, and a pencil

Procedure:

1. Gather the following materials: eight AA batteries of one particular brand; eight AA batteries of a second brand; and eight pen-size flashlights, or penlights, all of the same brand and size, each requiring two AA batteries.

2. Place two batteries of each brand in a zipper-lock bag, then put it in the freezer.

3. Place two batteries of each brand in a zipper-lock bag, then put it in the refrigerator.

4. Place two batteries of each brand in a zipper-lock bag, then put it in a room that maintains a moderate temperature.

5. Place two batteries of each brand in a zipper-lock bag, then put it in a warm place. (In front of a window that gets a lot of light is a good place.)

6. Wait three weeks. During this time, label four of the penlights with one of the battery brand names. Label the other four penlights with the other battery brand name. Then take four penlights labeled with one brand name and label one "freezer," one "refrigerator," one "room temperature," and one "warm." Label the remaining four penlights similarly.

7. When the three weeks have passed, put the batteries into the correctly labeled penlights. Then tape a 4-meter strip of newsprint along the bottom of a wall. Lay the penlights in a row and turn them on so that they shine a circle of light on the newsprint. You may need to move the penlights closer to or farther from the newsprint to have circles that fit on the paper.

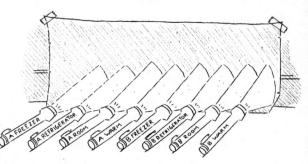

8. For three days in a row, check the penlights every hour for five hours, then trace the circle of light they shine on the newsprint, recording the time. Continue doing this until the lights no longer shine and all batteries are dead.

9. Compare your results.

Psychology: the study of the behavior of people.

Sociology: the study of organized groups of people.

Sources

Local

Look in your phone book for psychologists, psychiatrists, and universities (especially with professors that teach psychology).

National

American Psychological Association
750 1st St. NE
Washington, DC 20002
(202) 682-6000 ☛ Web site: **http://www.apa.org**

American Sociological Association
1722 N St. NW
Washington, DC 20036
(202) 336-5500 ☛ Web site: **http://www.asanet.org**

Science Fair Project
Who Knows What We Fear?

Many scientists use data collected in surveys. In this project, you survey girls and boys about their fears. It is very important to protect the identity of each subject surveyed and to keep the individual results confidential.

Background:

Everyone has fears, but are these different? Do we know what other people fear?

Hypothesis:

If I ask girls and boys to list their fears and also to list what they think others fear, then I think I will discover everyone has some similar fears. I'll also find that girls and boys understand their own gender's fears better than the other gender's fears.

Materials:

- 20 volunteers: 10 girls, 10 boys
- 20 3-by-5-inch note cards
- 20 pens
- 20 three-page surveys, stapled
- paper for making charts and lists (at least 10 pages)

Procedure:

1. Ask 10 girls and 10 boys to take a survey for you. Do not tell them in advance what it is about.

2. Number twenty 3-by-5-inch note cards 1 through 20. Under each number, write "Name." Leave enough space for a person to write a name there. Under that, write "Gender." Leave enough space for a person to write "Boy" or "Girl."

3. Make your survey. On the first page, write "List all your fears." On the second page, write "List all the fears girls have." On the third page, write "List all the fears boys have." Make 20 copies of your survey, and staple each copy.

4. Before you hand out the surveys, have each volunteer fill out a note card and return it to you. Then give each a survey, and instruct the volunteers to write on it their number from the note card, not their

name. Ask them to fill out the survey without talking to anyone and to give it back to you as soon as they are done.

5. Go through your note cards, then make a chart with two columns. In the first column, write "Girls," and in the second column, write "Boys." Then write the girls' numbers under the first column and the boys' numbers under the second column.

Girls #		Boys #	
1	16	3	17
2	18	4	19
5		6	
8	20	7	
10		9	
12		11	
13		14	
		15	

6. When you have the completed surveys back, make a pile of the girls' surveys, checking off the numbers from your chart. Check that the remaining surveys are boys'.

7. Take the first page of the girls' surveys, and compile a list of the fears the girls had written, calling it "Girls' Fears"—A. Next, take the second page of the girls' surveys, and compile another list of all the fears, calling it "Girls Predict Girls"—B. Then take the third page of the girls' surveys, and compile one more list of fears, called "Girls Predict Boys"—C.

8. Make three lists, as you did in step 7, with the boys' surveys, but call them "Boys' Fears"—D; "Boys Predict Girls"—E; and "Boys Predict Boys"—F.

9. Now compare the fears listed in A with B and E, and then compare the fears listed in D with C and F.

MATERIALS AND SUPPLIES

The materials you use in your project may vary from something as easy to find as a plastic bag to something as hard to locate as a vacuum jar. Remember to make your materials list clear and accurate so that anyone could assemble and use the exact same kind of materials you did.

After you have outlined the steps to your procedure, you need to list all the materials you will use in your project, just like a recipe lists all the ingredients required.

There are lots of places to get your supplies. First look at home, and ask your parents if they have what you need. Then you may want to shop at hardware stores, home and garden centers, pet stores, electronics stores, computer stores, pharmacies, medical supply stores, art supply stores, and stationery stores. Ask your teachers at school about school-owned supplies and equipment you may use. Your science teacher may receive catalogs from scientific supply companies you can order from, and your industrial shop teacher may be able to help you build some things you need. Local businesses often like to support educational experiences by providing equipment and information, especially when the project has something to do with the business itself. Sometimes industries or research centers may be interested in your data and results and will let you use laboratory facilities in exchange for sharing what you find.

KIDSOURCE TIPS

If you plan to order supplies from a catalog, remember to allow for the time it takes to first receive the catalog. (You may have to order the catalog first!)

Have backup supplies for things you will use up or that may break.

You may need some special scientific equipment that is hard to locate. Look for local scientific supply stores in your yellow pages under "Scientific Apparatus and Instruments." Also, you can order a student supply catalog from Carolina Biological Supply Company, although only your teacher can order from the general catalog, which has a greater variety of supplies.

Carolina Biological Supply Company
2700 York Rd.
Burlington, NC 27215
1-800-334-5551 ☛ Web site: **http://www.carolina.com**

SUPPLIES FOR EXPERIMENTS

You will need pens and notebooks, and perhaps a camera and film, for all experiments. Here is a general supply breakdown by the different sciences.

Astronomy supplies:

- binoculars, rulers, compasses, star charts, telescopes, cameras

Where you can find them: camera shops, hardware stores, home and garden centers

Biology supplies:

- pet food, pet cages, soil, seeds, plants, plant food, fertilizer, jars, lids, microscopes, tape measures, scales, nets, eyedroppers

Where you can find them: pet shops, home and garden centers, hardware stores, Carolina Biological Supply Company catalog

Chemistry supplies:

- chemicals, balances, hot plates, test tubes, graduated cylinders, racks, scales, thermometers, tongs, eyedroppers

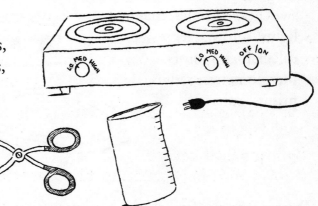

Where you can find them: chemical supply stores, pharmacies, home and garden centers, electronics stores, Carolina Biological Supply Company catalog

Earth sciences supplies:

- maps, weather charts, tidal charts, barometers (for measuring air pressure), thermometers, wind socks, plastic jars, calculators, picks, hammers, scales, tape

Where you can find them: hardware stores, home and garden centers, rock shops, map stores, marinas, Carolina Biological Supply Company catalog, U.S. Geological Survey (for maps)

Physics supplies:

- weights, pulleys, belts, springs, switches, batteries, magnets, lights, mirrors, lenses, calculators, motors, thermometers, radios, prisms, rulers

Where you can find them: hardware stores, camera stores, home and garden centers, computer stores, electronics stores, Carolina Biological Supply Company catalog

SUPPLIES FOR BUILDING YOUR EXHIBIT

Check pages 94 through 96 for instructions on building your exhibit. Stationery stores, art supply stores, and electronics and hardware stores are the best places to look for materials to construct an exhibit. Here are some materials you may need:

Art Supply Stores and Stationery Stores

- cardboard
- strong tape
- markers
- glue
- stick-on lettering and stencils
- scissors
- paint
- glitter
- clay
- construction paper
- stapler

Electronics and Hardware Stores

- Peg-Board (thin pressed-wood sheet with rows of holes for hanging pegs)
- foam-backed cardboard
- hinges, screws, nails
- lightbulbs, fixtures, hot element shields
- extension cords
- dowels
- plastic sheets
- putty
- plaster of paris
- strong picture wire
- hexagonal light mesh wire (chicken wire)
- aluminum foil
- hammer
- screwdriver
- wire cutter
- wrench
- stapler

KIDSOURCE TIPS

It's never too early to think about how much time you will need to get supplies, construct things, grow plants, and run your experiments. Here's a checklist of questions to help you prepare:

- Do you need to order any special materials or equipment?
- Do you, or does your teacher, have the catalogs you need to order these supplies from?
- Are you going to build anything that may be complicated and take time?
- Will you need to develop film?
- How long do seeds need to **germinate**? (If you are growing plants from seeds, look at the "days to germination" panel on the back of the seed packet, and choose seeds with a short germination period.)

START YOUR PROJECT!

"I'm way excited about my project this year. Last year I took water samples from the pond near my house. I found signs of pollution, and this year I want to take more samples and see what's changed. I hope the pollution is better, but you never know."

—Bettina Kwan, age 12

This is it—now you can actually start your project! If you haven't felt like a scientist yet, you are about to. You have been very busy planning and organizing your science fair project. Up to this point, you have

1. *found a project topic and thought of a question you want to answer.*
2. *researched your topic.*
3. *developed a hypothesis.*
4. *decided what your procedure will be and what materials you will need.*
5. *received permission to proceed if you are using tissue samples or are using animal or human subjects.*
6. *gathered your supplies.*

Now let the science begin!

BEGIN YOUR NOTEBOOK

One of the most important parts of your project is your notebook. Your notebook tells the story of your project from your very first idea to the end—it's a valuable record! As long as you keep your records up-to-date, you can always refer to your notebook for clues when something happens you don't expect or don't understand. Begin with the background of your project, and identify what interested you in your project in the first place. Write down your initial questions. Briefly review your research findings, and explain how you came to design your project, describing its stages of development—all the changes you made and the reasons for them. State your project question, your hypothesis, and your procedure; then list your materials.

Next, create a section where you will record your quantitative data, which is made up of numbers. You need a place where you can neatly record the measurements you make. Try to figure out the best way to record your quantitative data—will a table work? How many rows and columns will you need? How many subjects will you have? How many trials will you do? How many measurements will you make? What labels will you use? These are all things you need to consider when you are creating a place to record your quantitative data.

You also need to create a section for recording your qualitative data, where you will note how things happen as you do your experiment. You will have a lot of important qualitative data, because in every experiment you need to watch for more than just your expected responding variable. Other effects of your changing variable may also appear. Your qualitative data can alert you to interesting and unexpected discoveries that you may want to explore in other experiments. One good way to record your qualitative data is to make a list of the steps of your procedure, leaving a large blank space after each step for writing down your observations.

BUILD YOUR PROJECT

Now you will put together the physical part of your project. Some procedures call for a lot of building, but others don't. You should have an adult help you with this.

If you are doing field work, you may need to build a bug catcher or a sieve or to mark off the area you are studying with a simple wire fence. Perhaps you are building weather instruments or a telescope. Be sure that you understand how to use all tools and materials properly and that you put them away when you are done.

If you are doing an experiment, you may need to build some part of it and assemble and organize the rest of your materials. If you need to build any kind of structure, you may want to get expert advice from your shop teacher or from someone at a lumber or hardware store.

Do you have a safe place for your experiment, where no pets or family members will bother it? Be sure, too, to conduct your experiment in a place where it will not bother others. You may need to sterilize some things. Pots and containers can be soaked in a bleach solution to sterilize them or, if they will not melt, boiled in water. Soil can be baked on

That Covers It!

Here is an engineering experiment you can do to determine if round manhole covers are the best shape for the job. Have you ever wondered why manhole covers are round? They rest on a lip that is smaller than the cover—which leaves a space with a diameter smaller than the cover, so the cover can't fall in. Try building your own manhole models out of cardboard, and design other cover shapes to determine which works best. Can you figure out why some are better than others?

KIDSOURCE TIPS

Record your data with a pen, because pencil marks could wear away and become hard to read.
- Do not jot data down somewhere else to enter into your notebook later.
- Do not try to memorize data to write down later.

a tray in the oven at 400 degrees Fahrenheit to kill unwanted organisms. Do not place plants or animals near hot or cold drafts, windows, or vents. Be sure nothing is in direct sunlight unless that is your purpose.

Now you have gathered and constructed everything you need and have set up your experiment or are ready to begin your field work. So start your project!

RECORD YOUR DATA

Watch carefully as your experiment progresses and take careful measurements. Have your notebook with you and your charts or tables ready for entering data. When you are gathering data, you do not want to be distracted by anything else, and you want to record your observations immediately. Even when the data seem strange and are not what you expect, write down exactly what you see. Never change a piece of data because you do not think it is right.

PHOTOGRAPHS

Pictures tell a thousand words, and photographs are a great way to record results. Pictures are also good for making your science fair project display more interesting, and for showing equipment that is too big to bring to the fair. Everyone is interested in seeing how you did your project, and some results may be very dramatic in photographs. However, do not replace quantitative data with photographs—use them together. If you do use photography, be sure to allow plenty of time to get the film developed.

KIDSOURCE TIPS

You may want to blow up your best photos to 8 inches by 10 inches or larger for display in your exhibit. This is expensive, but you do get high-quality images. Or you may take them to a place that makes color photocopies. This runs about $2 to $4 per picture and can be done fairly quickly.

KEEPING OBJECTIVE

As you do your project, you want to be as **objective** as possible. That means you do an experiment or your field work with an open mind and document things as precisely as you can. You must be very careful that you do not look for only what you want to see but that you are alert to what really happens. When you design your experiment, you want to be very careful not to set it up so that you exclude real possibilities. In other words, your subjects must be free to react as they naturally do and should not be "helped" to react against their inclination.

DEALING WITH PROBLEMS

No matter how carefully you plan, you may still encounter problems. What should you do?

First of all, determine what kind of problem you have. Ask yourself these questions: Is your equipment running properly? Does something leak or overflow or not fit properly? Is your problem with your materials? Do you think you may have overdone something or used too much or too little of some material? Perhaps you did not find out the correct reaction or growth time in your research. Is something taking longer than you expected, requiring more time than you have? Are things moving too fast for you to measure them? Or is there a problem with the health or feeding of plants or animals?

Sometimes a simple adjustment can make the difference and put you back on track. Sometimes you will need to rebuild or replace a portion of your experiment and start over. In your notebook, write down your observations and any changes you make in your procedure, and then keep your procedure consistent for the rest of your experiment.

KIDSOURCE TIPS

Remember, the point of an experiment is not to prove something but to test something in order to learn something. A scientist never knows for sure what he or she will discover, and that is what makes science exciting!

Sometimes something is seriously wrong, and you need help. Perhaps you mixed up your labels, or perhaps you received mislabeled supplies or the wrong materials from your source. Your teacher can help you look for these things. But sometimes your experiment just doesn't work at all, or you have made such a big planning error that you are not able to see any reasonable results. If this happens and you still have enough time, you may want to think up a whole new experiment.

You may be getting unexpected results. That does not necessarily mean you have a problem. Because you have documented your procedure correctly, so you or anyone else can duplicate your results, you can check yourself. When you rerun your experiment for your second and third trials, each time you should get similar results. If your results are different each trial, you probably made an error somewhere along the way. Try to figure out what that error might be, correct it, and then rerun your experiment. If your results are similar each trial though still not what you expected, you may have overlooked a factor affecting your experiment. Try to think what that factor could be. It is possible you have made a discovery!

SOLVING PROBLEMS TO WIN!

Seventh-grader Zephyr Detrano won first prize at the Los Angeles County Science Fair in 1996 with her project, "Diffusion Rates in Liquids." Zephyr was studying how different substances mix with each other and had some problems to iron out in the beginning.

For one thing, Zephyr's test tubes were too confined to show the diffusion, making it difficult for her to monitor her experiment. So she tracked down Erlenmeyer flasks, which have a wider, flattened bottom. Then one of the substances she was testing, isopropyl alcohol, took a very long time to diffuse.

Zephyr had a theory to explain what happened, but she didn't have the time to wait for it. She learned a lot: "It was interesting to see that it took longer for the molecules that were heavier to diffuse. In future research I would try gases, different liquids, and liquids at different temperatures."

What if nothing really happens? Sometimes your experiment does not show a change at all, and you have **inconclusive results**. Inconclusive results, if you have not made an error, are still valid results. Perhaps you thought that by altering the wing shape, you could make a faster flying paper airplane but were disappointed to find that it did not fly faster at all. Your experiment is valid; it just showed that your hypothesis is not. Your experiment is still important, because if you were to publish your results, as scientists do, other scientists would not need to try your wing design, since they would now know the result. Your experiment could save other scientists a lot of trouble and guide them to look at other possibilities.

When the results you get from a carefully performed experiment prove your hypothesis wrong, your project is a valid contribution to science. You (and others) have a chance to think about the validity of the assumptions you based your hypothesis on. In your conclusion, you should try to analyze what happened and suggest ways to explore the result further. Remember, not every hypothesis is correct—that is why scientists test them!

YOUR CONCLUSION

Now you have started your notebook, built your project, run your experiment or done your field work, and recorded your data. It's time to come to a conclusion!

If you did field work:

- Describe any noticeable patterns in the data you collected or the behavior you observed, and suggest an explanation for them.

- Critique your project and suggest improvements. How could you have been more thorough and done a better job? Did you overlook anything?

- Discuss what else you would like to study, based on the knowledge you gained from doing your project and on additional questions that came up.

If you did an experiment:

- Discuss whether your hypothesis was proved right or wrong. If it was right, this discovery should lead you to new questions to explore. If your hypothesis was wrong, or if your results were inconclusive, you will discuss why, based on what you think happened.

- Discuss any errors you may have made and how to avoid them in the future. How could you have done your experiment better? How could you have been more careful, more thorough? Were there any variables for which you did not allow?

- Discuss what you would change if you did this experiment again and what else you would like to find out.

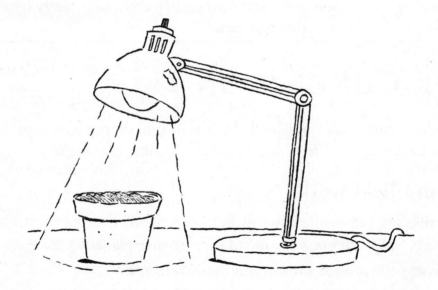

KIDSOURCE TIPS

So you find the days and weeks have flown by, and you are asking yourself, "How am I going to get this project done?" Never fear. Below you will find 10 timesaving ideas to get your project to the fair on time!

1. If you are using photography, find an overnight developer or one-hour photo shop.

2. Soak large seeds like corn, peas, and beans the night before you plant them to help them sprout faster.

3. Keep soil warm with lights (unless you are testing the effects of temperature) to help seedlings start more quickly.

4. Wisconsin fast plant (Brassica rapa) is a quick-growing plant. You can order it from Carolina Biological Supply Company, 2700 York Road, Burlington, NC 27215-3398.

5. Buy extras of all your supplies, so that you aren't stuck at the last minute because you need something you used up, broke, or lost.

6. Try to organize your experiment so that you can run all your trials at the same time, instead of one after the other.

7. Have your science teacher and your parents check for problems with your experimental design before you begin.

8. Get permission ahead of time to use professional facilities or go to places, and make sure they are open and available when you plan to use them.

9. Many projects are usually done in the winter, when daylight is short and weakest, so you may want to buy grow lights for plants to help them grow faster.

10. Check schedules of natural events that are important to your experiment, such as tides; the moon's phases; meteor showers; egg hatchings; flower bloomings; rainy seasons; and animal denning, hibernation, migration, rutting, mating, and birthing times.

GET READY FOR THE FAIR!

You worked hard—now show it off! It's time to write your report, build a dynamite display, and head for the fair!

WRITING YOUR REPORT

If you took good notes throughout your research, experiment, or field work, writing your report will be a breeze. Use a computer, word processor, or typewriter to make your report look neat and easy to read. Your report should contain

- a title
- a description of the background of your project
- your hypothesis
- your procedure
- a list of materials
- your results
- your conclusion
- your bibliography.

Think of a title for your project. Your title should indicate what is being tested or observed. Try to think of a title that's catchy *and* explains what you did. Judges love that! For instance, if you did an experiment to determine if seeds grow faster on clear, sunny days, you could call it "At Light Speed."

Next, you need to write a paragraph or one-page description of the background of your project—what got you interested in it?

Then you state your hypothesis.

After that, detail your procedure: the steps you took, the measurements you made, how you collected your information and recorded your observations. Explain how your equipment worked. If you had to build any special equipment, specify the exact directions, measurements, and supplies you used. Identify and describe the parts of your experiment: your control and experimental groups, your changing and responding variables.

Next, list your materials. Be specific, and note the exact amounts of the materials you used. If you used plants that were 6 inches tall, say so. If you used 1 liter of water, include that information.

Then show your results. Explain any difficulties you encountered. Describe how you made adjustments and why you thought they were necessary. Any graphs, charts, or tables of data that you create belong in this section.

Now comes your conclusion, which is your summary of what happened, based on your results. Look for patterns in your data and explain them. Determine whether your data support your hypothesis. If not, try to explain why. Evaluate your project. Suggest improvements. If you had to start over, what might you do differently and why? If you could continue this project or one related to it, how would you go about that? Are there any factors you had not previously considered that could have accounted for your results? What did you learn from your project?

Finally, your bibliography goes at the end of your report. Also list and thank any experts who may have helped you. Describe any work you did not do yourself, and acknowledge who did do it—for instance, give the name of the laboratory you sent samples to for special analysis.

KIDSOURCE TIPS

- When you buy your supplies to make your display, get extra of everything, especially letters. You don't want to run out of *E*s when the stores are closed.
- Keep your display clean and tidy in a safe place until you set it up at the fair.
- Bring a little emergency kit with you—touch-up paint, markers, glue, tape, extra lettering, and spare parts like hinges and screws, batteries, bulbs, and extension cords—just in case you need anything at the last minute.

When you finish your report, you need to proofread it. Look for misspellings, typos, run-on sentences, and so on. Then give it to others to read. Ask them if all sentences are clear and easy to understand. They may find misspellings and grammatical errors you missed. Correct any errors, and if you used a computer, print out a clean copy of your report.

BUILDING YOUR EXHIBIT

If your science project were a picture, your exhibit would be the frame. You use your exhibit to hang printouts of your report and illustrations of your science fair project. Be sure you know what the allowable size is before you begin to build your display. Your exhibit needs to be able to stand on its own, so a two- or three-sided display works best. Usually displays are built from cardboard, masonite, plywood, or Peg-Board.

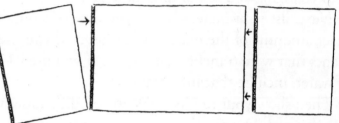

If you are using cardboard, take two pieces that are about 2 feet by 3 feet, and cut one piece in half lengthwise, creating two 3-by-1-foot pieces. Connect each of the two halves to opposite sides of the bigger piece. You can connect the sides with strong wide tape. This will give your exhibit a self-standing background.

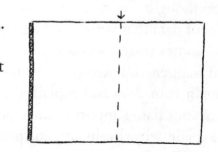

HINGE WITH STRONG TAPE

If you use wood or Peg-Board, you will need to have them cut to the dimensions you want—the most popular dimensions are a 2-by-3-foot back piece, and two 1-by-3-foot side pieces. Hinges will hold the pieces together best.

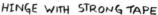
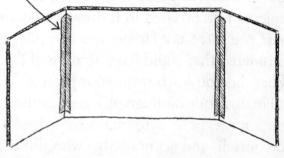

Use your report for the text you put on your display. Print out on separate sheets your

- background information
- hypothesis
- procedure

- materials list
- results
- conclusion.

Your exhibit should read from left to right. Arrange your text on your display like this:

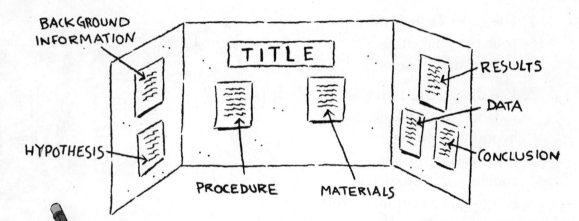

Neon Lights

Seventh-grader Brian Bourget won first prize in the Earth Sciences Division at the Los Angeles County Science Fair in 1996 for his project, "How Does Ultraviolet Light Affect Minerals?" Here's how his prize-winning project was born: "I'd always had a thing for rocks, and then my uncle told me about fluorescent rocks, which I thought were pretty cool. I wondered how and why they fluoresced." Brian's exhibit was cleverly decorated with neon-colored paper, which obviously caught the judges' attention.

1. your title on the top of the middle of the center panel

2. your procedure and materials on the center panel, below title

3. your background information and your hypothesis on the left panel

4. your results with data, plus conclusion and evaluation, on the right panel

Leave a copy of your full report with its bibliography, along with your notebook, on the table in front of your exhibit. It is a good idea to attach both of these to your display with a long string, so they don't get separated and lost.

DISPLAY TIPS

To make your project stand out and get the judges' attention, try these "eye-catching" tips:

1. **Use extra-large lettering** for your title, big enough to be seen from a distance.

2. **Use distinctive lettering** to label your background, hypothesis, procedure, materials, results, and conclusion. Buy stick-on lettering at a stationery store to more easily make your display look neat.

3. **Choose colored cardboard** to save the time it would take to paint your display.

4. **Use colored backgrounds or special matting** for each section on your display board. Paint decorative borders or illustrations, and make frames for the pages of your report and the photographs you use.

5. **Try to use charts and graphs**—they are eye-catching and help the viewer read results easily.

6. **Be artistic**—use great drawings, photographs, charts, and graphs. Include colorful touches: borders, paint, glitter, foil, colored glue.

7. **Avoid clutter**—organize your board so that it tells a story, with your eye being led from left to right as the story unfolds. Keep your display neat.

8. **Be creative**—if you have a bar graph showing rainy days per month, use lightning as the bars.

9. **Display any equipment
   you used** if it will fit in
   front of the center
   panel. Showing what
   you actually did will grab
   attention. When your
   equipment is too large to
   display, interesting photographs on the center panel
   will work OK.

THE JUDGES

The judges are a group of teachers, scientists, and
professionals who enjoy
encouraging students in
science. They have
looked at your display
and read your report
and are looking
forward to talking
with you. They want
to see that you
understand your

background research, know the scientific method, and have done your own
work. Here are some of the categories they will be giving points for:

CREATIVITY

- The problem or the approach to the problem is original.
- Equipment and materials are used in a clever way.
- The student is aware of and interested in the unanswered
  questions that remain.
- The project has a clear purpose.

SCIENTIFIC THOUGHT

- The hypothesis is clearly stated.
- The project shows depth of study and effort.
- The data collected are appropriate for the problem.
- Scientific procedures are accurate and organized.
- Conclusions are logical, based on the data collected, and relevant to the hypothesis.

THOROUGHNESS

- The scientific literature has been searched.
- Experiments have been repeated.
- Careful records have been kept.

SKILL

- The project is properly designed to give valid, reliable, and accurate data.
- Special observational, mathematical, computational, carpentry, or technical skills are shown.

CLARITY

- The project notebook is well organized, neat, and accurate.
- The report is complete with bibliography.
- The title accurately reflects the problem.
- The purpose, procedures, and conclusions are clearly outlined.

ORAL ABILITY

- The student can talk knowledgeably on the subject of his or her project.
- The student understands and can explain the parts of a controlled experiment, a field study, and the scientific method.
- The student can discuss the ideas and principles of his or her project.
- The student is able to make predictions based on the results of his or her experiment.

TALKING TO THE JUDGES

You are ready to present your project to the judges. You are on time, standing by your project, dressed neatly. Be prepared to courteously answer the judges' questions. They will ask you:

1. What is the purpose of your experiment?

2. Why did you choose your project?

3. How did you do your project?

4. What was your conclusion?

5. When did you start your project?

6. How did you build your equipment?

7. How did this device work?

8. Where did you get your supplies?

9. Did you have any problems with your project?

10. Did you discover any problems?

11. How did you handle any problems?

12. What did you learn from your project?

13. What would you do differently?

14. What else would you like to do to explore your topic subject?

15. Can you tell us about your school and your hobbies?

Practice answering these questions and talking in a relaxed way about your project in front of friends and family, so you will feel more comfortable talking to the judges.

A SAMPLE INTERVIEW WITH THE JUDGES

Here is part of an interview to show you what it may be like. You can act this out with friends and family for practice.

JUDGE 1   Hi. I'm Ms. Kelay, and this is Ms. Green, Mr. Peal, and Ms. Bond.

YOU   Hello. I'm (your name).

JUDGE 2   How did you think of your project?

YOU   I love hiking, and on my favorite hike I see a lot of dragonflies. I decided to make a collection to see how many kinds there were. I discovered five local species, and I identified a new one with the help of Dr. Siegal at the Natural History Museum.

JUDGE 3   That sounds exciting. How did you collect the dragonflies?

YOU   I went to the Sleeping Bear Valley every day for a week. I used a net to sweep the area around a pond.

JUDGE 4   If you did your project over again, what would you do differently?

YOU   I would like to collect the dragonflies in all their stages of development.

JUDGE 2   That sounds interesting.

YOU   I think so. I'm going to do that next year.

JUDGES   You've done a great job, and it looks like you had a lot of fun.

YOU   Thank you. Yes, I did.

JUDGES   Our time is up. Good-bye. It was nice meeting you.

YOU   Good-bye. It was nice meeting all of you, too.

CLEANING UP

It's not over yet! You must take your exhibit apart and bring it home. Clean up any mess you may have left from your display.

WHAT'S NEXT?

Perhaps you are already thinking of next year's project. Many prize-winning entries are projects suggested by an earlier project. If you liked your topic and want to learn more about it, you may already be planning your next project. And it can be an especially good one if you start early.

You may want to enter more science fairs, and try to win your way to international fairs, where the prizes are jobs, scholarships, and scientific recognition. You may want to find out about other kinds of science projects, which are not done for science fairs but offer interesting experiences, like designing experiments that could possibly be done on a U.S. space shuttle. Your science projects can take you to fascinating worlds of discovery: to international exchange programs and to working on teams to solve community, state, national, and international problems. This is just the beginning. There is a whole world to explore.

Here are some other organizations you can write to and find out what they offer:

Westinghouse Talent Search
Science Service
1719 N St. NW
Washington, DC 20036
(202) 785-2255

The National Junior Science and
Humanities Symposium
JSHS National Office
Academy of Applied Science
98 Washington St.
Concord, NH 03301

THE TOOLS OF SCIENCE

When scientists measure something, they use the metric system. Metric units are easier to calculate with and easier to relate to one another than are non-metric units, such as inches and feet or pints and gallons. You may be used to the U.S. conventional measuring system. Check out these charts if you are unsure what unit to use or want to convert your measurements.

UNITS OF MEASUREMENT

Length
kilometer (km)
meter (m)
centimeter (cm)
millimeter (mm)

Area
square centimeter (cm^2)

Volume
liter (l)
milliliter (ml)

Concentration
mole (mol)

Mass
gram (g)

Power
watt (W)

Energy
calorie (cal)
joule (J)

Frequency
hertz (Hz)

Electrical Units
volt (V)
ampere (A)
ohm (Ω)

Temperature
degrees Celsius (ºC)

Time
hour (hr.)
minute (min.)
second (sec.)

METRIC UNITS

1 liter (l) = 1,000 milliliters (ml)

1,000 liters (l) = 1 kiloliter (kl)

1 gram (g) = 1,000 milligrams (mg)

1,000 grams (g) = 1 kilogram (kg)

1,000 kilograms (kg) = 1 metric ton (t)

1 meter (m) = 100 centimeters (cm)

1 meter (m) = 1,000 millimeters (mm)

1 kilometer (km) = 1,000 meters (m)

CONVENTIONAL U.S. UNITS

12 inches (in.) = 1 foot (ft.)

3 feet (ft.) = 1 yard (yd.)

1,760 yards (yd.) = 1 mile (mi.)

3 teaspoons (tsp.) = 1 tablespoon (tbsp.)

1 ounce (oz.) = 2 tablespoons (tbsp.) = 6 teaspoons (tsp.)

1 cup (c.) = 16 tablespoons (tbsp.) = 8 ounces (oz.)

1 pint (pt.) = 2 cups (c.) = 16 ounces (oz.)

1 quart (qt.) = 2 pints (pt.) = 4 cups (c.) = 32 ounces (oz.)

1 gallon (gal.) = 4 quarts (qt.) = 8 pints (pt.) = 16 cups (c.) = 128 ounces (oz.)

1 pound (lb.) = 16 ounces (oz.)

CONVERSIONS

Here are some conversions for temperature:

To convert degrees F (Fahrenheit) to degrees C (Celsius), use this formula by plugging in your Fahrenheit number, subtracting 32, and then multiplying by ⅝:

$$C = (F - 32) \tfrac{5}{9}$$

To convert degrees C (Celsius) to degrees F (Fahrenheit), use this formula, plugging in your Celsius number, multiplying by ⅑, and then adding 32:

$$F = (\tfrac{9}{5} C) + 32$$

CONVERTING TO METRIC

IF YOU NEED:	BUT YOU HAVE:	MULTIPLY BY:
centimeters	inches	2.54
meters	feet	0.305
kilograms	pounds	0.45
milliliters	ounces	30.0
milliliters	tablespoons	14.8
milliliters	teaspoons	4.9
milliliters	cups	236.5
liters	pints	0.47
liters	quarts	0.95
liters	gallons	3.78

CONVERTING TO U.S. CONVENTIONAL MEASUREMENTS

IF YOU NEED:	BUT YOU HAVE:	MULTIPLY BY:
inches	centimeters	0.394
feet	meters	3.281
pounds	kilograms	2.205
ounces	milliliters	0.034
tablespoons	milliliters	15
teaspoons	milliliters	5
cups	milliliters	240
pints	liters	2.113
quarts	liters	1.057
gallons	liters	0.264

SIGNIFICANT DIGITS

When you are multiplying or dividing numbers that come from measurements you have made, you need to know about **significant digits**. First count the places of your number. A place is the number of digits after a decimal point. For instance, 1.0 has one place, 1.20 has two, and 1.440 has three. Remember how, when you are measuring, the number in the last place is estimated? That is the number you get when you estimate the distance to the nearest line of the smallest mark on your standard measuring device. This number takes the last place position of your significant digits. You must always keep the same number of places when you use numbers you have measured in calculations. Otherwise, it will seem that you have measured more or less exactly than you actually have.

What kind of calculations could give you extra places? Well, multiplication and division could. When you are calculating area or volume or converting to a metric number, you might multiply your measured number and get one with more places. Simply round up or down to keep the right number of significant digits. Round up to the next highest number in the place before the last number when the last number is five or greater. Round down, or simply drop the last number, when the last number is less than five.

For example, if you were measuring the sides of a rectangle and got 5.6 centimeters for one side and 8.9 centimeters for another, you have one place in your measured numbers. When you multiply your numbers to get the area of the rectangle, you get 49.84 square centimeters. But this number has two places, which means that it measures one place more accurately than your device actually did, which we know is not right. So you would round down to 49.8 centimeters.

TIMETABLE

Use this timetable as a model to plan the organization of your science fair project. You must understand before you start how much time it will take to do the things you plan to do, such as develop film, grow plants or seeds, or watch animals develop. You must leave time to change things or even start over, if necessary. This timetable gives you four months, but you can add to that if you want to start earlier and spend more time on some things. Remember that most of your work will have to be done on weekends and after school.

WEEK 1

❏ Follow the questions in this book to choose your topic.

❏ Order your scientific supplies catalog.

WEEK 2 – 4

❏ Go to the library and contact experts to research your topic.

❏ Create your bibliography.

WEEK 5 – 7

❑ Plan your project. Fill out and get forms OK'd to use tissue samples or animal or human subjects, if that is allowed at your fair.

❑ Collect the materials you will need to do your project.

WEEK 8 – 10

❑ Start your notebook and run your experiment or do your field work. If you are using photography, develop the film.

WEEK 11 – 13

❑ Continue your experiment or field work. Redo your experiment or field work, if necessary.

❑ And if you have time, begin writing your report and building your display.

WEEK 14 – 16

❑ Write and proofread your report.

❑ Get supplies to create your display and put together your exhibit.

WEEK 17

❑ Set up your exhibit and enjoy the science fair!

FOR FURTHER READING

BOOKS

Adair, Robert K. *The Physics of Baseball*. New York: Harper & Row, 1990.

Amato, Carol. *Super Science Fair Projects*. Los Angeles: Lowell House Juvenile, 1994.

Barnes-Svarney, Patricia. *The New York Times Science Desk Reference*. New York: Macmillan, 1995.

Barr, George. *Sport Science for Young People*. New York: Dover, 1990.

Beller, Joel. *So You Want to Do a Science Project!* New York: Arco, 1981.

Bochinski, Julianne Blair. *The Complete Handbook of Scientific Projects*. New York: John Wiley & Sons, 1996.

Cobb, Vick. *More Science Experiments You Can Eat*. New York: J.B. Lippincott, 1979.

DeVito, Alfred, and Gerald H. Krockover. *Creative Sciencing: Ideas and Activities for Teachers and Children*. 2d ed. 1980.

Dixon, Dougal, and Raymond L. Bernor. *The Practical Geologist*. New York: Simon & Schuster, 1992.

Gardner, Martin. *Entertaining Science Experiments with Everyday Objects*. New York: Dover, 1981.

Gardner, Robert. *Experimenting With Science in Sports*. New York: Franklin Watts, 1993.

Gutnik, Martin J. *How to Do a Science Project and Report*. New York: Franklin Watts, 1980.

Hyer, James E. *Mister Wizard's Supermarket Science*. New York: Random House, 1980.

Ruben, Gabriel. *Electricity Experiments for Children*. New York: Dover 1960.

Smolinski, Jill. *50 Nifty Super Science Fair Projects*. Los Angeles: Lowell House Juvenile, 1995.

Tocci, Salvatore. *How to Do a Science Fair Project*. New York: Franklin Watts, 1986.

MAGAZINES

Know Your World Extra
A children's magazine that focuses on natural history and science, for ages 8 to 15.
Field Publications
PO Box 16630
Columbus, OH 43216

Odyssey
A children's monthly magazine for ages 9 to 15.
1027 N. 7th St.
Milwaukee, WI 53233

Popular Science
An easy-to-read monthly magazine, with highlights of science news and new products and technologies.
PO Box 51286
Boulder, CO 80322

Science Weekly
A weekly children's magazine for ages 9 to 13.
Subscription Department
2141 Industrial Parkway
Silver Spring, MD 20904

Super Science Blue Edition
A children's science magazine for ages 9 to13.
Scholastic
PO Box 3710
Jefferson City, MO 65107-9957

3-2-1 Contact
A children's science and math magazine for ages 8 to 14.
PO Box 53051
Boulder, CO 80322-53051

GLOSSARY

arboretum: a place where plants and trees are grown for study.

astronomy: the study of objects in outer space.

bibliography: a list of reading material that gives information on a special topic.

biology: the study of all living things.

botany: the study of plants.

calibrated tool: a measuring tool with precisely marked gradations.

changing variable: the one thing altered in a subject's environment that will produce a particular change in the subject.

characteristic: a specific trait. In a science fair project, a characteristic of something is chosen to study.

chemistry: the study of the composition, structure, and changes in composition and structure of substances.

conclusion: the last section in a science fair project report that explains what happened in the experiment or field study.

control group: in an experiment, the group that is not exposed to the changing variable.

data: numerical information or results.

environment: the conditions that surround places or living things and determine how they exist.

experiment: observations made in a controlled, artificial environment.

experimental group: the group of experimental subjects that are exposed to a changing variable.

field work: the observation and description of a specific environment at a specific time.

fossil: a permanently preserved plant or animal in rock.

geology: the study of the composition and physical characteristics of the earth.

germinate: to start to grow.

hypothesis: an educated guess or prediction of what may happen in an experiment or field work.

inconclusive results: when an experiment has no results.

marina: a place where boats are docked in water.

materials: the tools needed for a science fair project.

meniscus: the curved upper surface of a volume of liquid.

mentor: someone who helps by teaching or coaching.

meteorology: the study of the weather and Earth's atmosphere.

microbiology: the study of living things too small to be seen with the naked eye.

microfilm: a film with miniature photographs, used for storing information, including newspapers and magazines.

micrometeorites: very small pieces of objects from outer space that fall to Earth's surface.

objective: to keep an open mind and not "look" for a certain answer.

observatory: a place with a telescope for studying the sky.

oceanography: the study of the world's oceans.

paleontology: the study of ancient life from fossil records.

periodical: a magazine or journal.

physics: the study of the physical properties, such as energy and motion, of things.

planetarium: an indoor theater that has images of the night sky projected on the walls and ceiling.

predict: to guess.

procedure: the steps in an experiment or in field work.

psychology: the study of the human mind and human behavior in individuals.

qualitative data: results you describe with words.

quantitative data: results you measure and describe with numbers.

report: the written matter in a science fair project that explains the entire process, from the hypothesis to the conclusion.

responding variable: in an experiment, the sign of change you measure in a test subject.

scientific method: the steps to investigation scientists use, which involve an observation, a hypothesis, an experiment or field work, and a conclusion.

significant digit: one of the digits in a number beginning with the digit farthest to the left that is not zero and ending with the last digit farthest to the right that is not zero or is a zero considered to be exact.

sociology: the study of people in organized groups.

species: a category of biological classification, below genus, of related organisms that can interbreed.

terrain: the physical features of a piece of land.

test: to perform an experiment to see if a hypothesis is correct.

tide pool: a temporary environment of pools of sea water trapped between shoreline rocks after high tide.

toxin: a poisonous substance.

trial: an experiment that is performed from beginning to end.

zoology: the study of animals in their native habitat.

INDEX